A User's Guide to Operator Algebras

CANADIAN MATHEMATICAL SOCIETY SERIES OF MONOGRAPHS AND ADVANCED TEXTS

Monographies et Études de la Société Mathématique du Canada

EDITORS: Jonathan M. Borwein and Peter B. Borwein

Frank H. Clarke **Optimization and Nonsmooth Analysis*

Erwin Klein and Anthony C. Thompson **Theory of Correspondences: Including Applications to Mathematical Economics*

I. Gohbert, P. Lancaster, and L. Rodman *Invariant Subspaces of Matrices with Applications*

Jonathan M. Borwein and Peter Borwein *Pi and the AGM—A Study in Analytic Number Theory and Computational Complexity*

John H. Berglund, Hugo D. Jünghenn, and Paul Milne **Analysis of Semigroups: Function Spaces Compactifications Representation*

Subhashis Nag *The Complex Analytic Theory of Teichmüller Spaces*

Manfred Kracht and Erwin Kreyszig **Methods of Complex Analysis in Partial Differential Equations with Applications*

Ernest J. Kani and Robert A. Smith *The Collected Papers of Hans Arnold Heilbronn*

Victor P. Snaith **Topological Methods in Galois Representation Theory*

Kalathoor Varadarajan *The Finiteness Obstruction of CTC Wall*

G. Watson **Statistics on Spheres*

F. A. Sherk, P. McMullen, A. Thompson, and A. Weiss *Kaleidoscopes: Selected Writings of H.S.M. Coxeter*

Peter A. Fillmore *A User's Guide to Operator Algebras*

Alf van der Poorten *Notes on Fermat's Last Theorem*

P. Lancaster and K. Salkauskas *Transform Methods in Applied Mathematics: An Introduction*

---

*Indicates an out-of-print title.

A User's Guide to Operator Algebras

Peter A. Fillmore
Dalhousie University
Halifax, Nova Scotia

A Wiley-Interscience Publication

JOHN WILEY & SONS, INC.

New York • Chichester • Brisbane • Toronto • Singapore

This text is printed on acid-free paper.

Published simultaneously in Canada.

Library of Congress Cataloging-in-Publication Data

Fillmore, Peter A., 1936–
A user's guide to operator algebras / Peter A. Fillmore.
p. cm. — (Canadian Mathematical Society series of monographs and advanced texts)
"A Wiley-Interscience publication."
Includes index.
ISBN 0-471-31135-9 (cloth : alk. paper)
1. Operator algebras. I. Title. II. Series.
QA326.F55 1996
512′.55—dc20 95-38340

Printed in the United States of America

10 9 8 7 6 5 4 3 2 1

To Anne Ellen

Contents

Preface

The subject of operator algebras, now more than fifty years old, has in recent years experienced tremendous growth and vitality, with significant applications to other areas both within mathematics and outside. The award of Fields Medals to Alain Connes (1982) and Vaughan Jones (1990) for their work on operator algebras gives an indication of the importance of these developments. The aim of this book is to make some of this excitement accessible to the non-specialist, as painlessly as possible.

To achieve this goal in a book of a size that does not unduly tax the reader's stamina has required certain compromises. Some important topics are not treated at all. An obvious example is cyclic cohomology (but in this case it would be difficult to improve on the presentation in [Con94]). A number of topics are discussed only briefly. It has been necessary to omit a number of proofs and to give only sketches of others. In making such choices the guiding principle has been to expose the main ideas and their interconnections as clearly as possible. For complete proofs appropriate sources are suggested. Illustrative examples are emphasized throughout the book.

The treatment presupposes a background of general functional analysis and operator theory through the spectral theorem and index theory. The first five chapters constitute a fairly comprehensive introduction to the basic theory of C^*-algebras and von Neumann algebras, including most proofs and many examples. That accomplished, the goal in the next four chapters is to introduce the work of Connes (and others) on amenable von Neumann algebras and of Jones on factors and knot theory. The final chapter, on K-theory, also includes a description of very recent work in the direction of classifying amenable C^*-algebras by means of K-theoretical invariants. Much of the material of the last five chapters involves highly technical details and few complete proofs are given, though a number are outlined. A longish bibliography (150 items) is included to assist the interested reader. More comprehensive bibliographies may be found in [Dix77] (2052 items), [Tak79] (417 items), [Bla86] (208 items) and [Con94] (599 items).

The book contains no formal exercises, but there any many opportunities for the reader to supply details or more substantial parts of arguments. It should be suitable as a textbook for a second year or higher graduate course. In fact much of this material was prepared for a graduate class given by me at Dalhousie University in the spring of 1992. I acknowledge with thanks the interest and helpful suggestions of my students and colleagues: Keith Johnson, Terry Quinn, Heydar Radjavi, Alex Simonič and Yong Zhong.

The manuscript was completed during the special year on operator algebras and applications held at the Fields Institute in 1994–95. I have benefited more than I can say from the stimulation provided by the lively group of students, post-doctoral fellows and senior visitors assembled for the year. I particularly wish to thank John Chadam and his Institute staff for the splendid working conditions, and George Elliott, who as Chairman of the scientific programme committee for the year deserves a major share of the credit for its success. I am indebted to George Elliott, Peter Friis, Alex Kumjian, Jamie Mingo, Gert Pederson, Chris Phillips, Mikael Rørdam, Larry Schweitzer and Wend Werner, among others, for helpful conversations and for comments on parts of the manuscript; they have corrected some of my misconceptions and have helped improve the exposition substantially.

Finally I wish to thank Maria-Fe Elder, who prepared the first LaTeX draft, and Jonathan Samuel and Claus Dahl, whose patient explanations of the intricacies of TeX enabled me to take care of subsequent drafts.

PETER A. FILLMORE

Halifax, Nova Scotia
February 1996

A User's Guide to Operator Algebras

Chapter One

Fundamentals

1.1 C^*-ALGEBRAS AND MORPHISMS

Let A be an algebra over the complex field **C**. An **involution** on A is a conjugate linear map $a \to a^*$ such that $a^{**} = a$ and $(ab)^* = b^*a^*$ for all $a, b \in A$. A norm on A is a **C^*-norm** if the **C^*-identity**

$$\|a^*a\| = \|a\|^2 \qquad \text{for all } a \in A$$

is satisfied. A **C^*-algebra** is a complex algebra with an involution and a C^*-norm in which it is complete. The C^*-identity is equivalent to the condition

$$\|a^*a\| \geq \|a\|^2 \qquad \text{for all } a \in A$$

in any normed algebra with involution. In fact

$$\|a\|^2 \leq \|a^*a\| \leq \|a^*\|\|a\|$$

so $\|a\| \leq \|a^*\|$; replacing a by a^* in this inequality shows that involution is an **isometric** map ($\|a^*\| = \|a\|$ for all a) and the C^*-identity follows by substituting this equation in the above expression.

If a nonzero C^*-algebra A has a **unit element** 1 ($1a = a1 = a$ for all $a \in A$), then necessarily $1^* = 1$ and $\|1\| = 1$.

The **morphisms** appropriate to C^*-algebras are the **$*$-homomorphisms**: algebra homomorphisms that respect the involutions. We will see later that such maps are automatically **contractive** ($\|\varphi(a)\| \leq \|a\|$ for all a) and are necessarily isometric ($\|\varphi(a)\| = \|a\|$ for all a) when injective. We will be interested in both the category of C^*-algebras and $*$-homomorphisms and the category of unital C^*-algebras and unital $*$-homomorphisms ($\varphi(1) = 1$). In both categories the **isomorphisms** are the bijective $*$-homomorphisms (i.e., $*$-isomorphisms).

1.1.1

The fundamental examples of commutative C^*-algebras are $\mathbf{C}$ and, more generally, the algebra $C_0(X)$ of continuous functions $X \to \mathbf{C}$ vanishing at infinity on a locally compact Hausdorff space X, with

$$f^*(x) = \overline{f(x)} \quad \text{and} \quad \|f\| = \|f\|_\infty = \max|f|$$

The basic examples of noncommutative C^*-algebras are the matrix algebras $M_n(\mathbf{C})$ and, more generally, the algebra $\mathcal{B}(\mathcal{H})$ of bounded linear operators on a Hilbert space $\mathcal{H}$. Here the involution and norm on $\mathcal{B}(\mathcal{H})$ (hence on $M_n(\mathbf{C})$ by taking $\mathcal{H} = \mathbf{C}^n$) are determined by the inner product and norm on $\mathcal{H}$ by the requirements

$$\langle T^*\xi, \eta\rangle = \langle \xi, T\eta\rangle \qquad \text{for all } \xi, \eta \in \mathcal{H}$$

$$\|T\| = \sup\{\|T\xi\| : \xi \in \mathcal{H} \text{ and } \|\xi\| \le 1\}$$

When $\mathcal{H}$ is infinite-dimensional $\mathcal{B}(\mathcal{H})$ is no longer simple: the algebra $\mathcal{K}(\mathcal{H})$ of **compact operators**, defined as the C^*-subalgebra of $\mathcal{B}(\mathcal{H})$ generated by the operators of finite rank, is a nontrivial closed ideal (the only one if $\mathcal{H}$ is separable).

1.1.2

Here are some examples of morphisms of C^*-algebras. For any C^*-algebra A the morphisms $\mathbf{C} \to A$ are the maps $\lambda \to \lambda e$, where $e \in A$ satisfies $e = e^2 = e^*$ (i.e., e is a projection). In general, there need not be any nonzero morphisms $A \to \mathbf{C}$. This is the case for $A = M_n(\mathbf{C}), n > 1$, since this algebra has no proper ideals. The morphisms $C_0(X) \to \mathbf{C}$ are just the evaluations $f \to f(x), x \in X$.

If X is compact and $\varphi : C(X) \to C_0(Y)$ is a morphism, then there is a compact open subset U of Y and a continuous map $\alpha : U \to X$ such that

$$(\varphi(f))(y) = \begin{cases} f(\alpha(y)), & y \in U \\ 0, & y \notin U \end{cases}$$

for all $f \in C(X)$. In fact U is just the support of the idempotent $\varphi(1)$. If X is only locally compact the same formula holds, except that U need not be compact but α must be **proper**, in the sense that $\alpha(u_\lambda) \to \infty$ for any net $u_\lambda \to \infty$ in U.

Nonzero morphisms $M_k(\mathbf{C}) \to M_n(\mathbf{C})$ exist only if $n \geq k$; then they are precisely the maps of the form

$$a \to U\,\mathrm{diag}(a, a, \ldots, a, 0)U^*$$

with $U \in M_n(\mathbf{C})$ unitary. Here the number of repetitions of a on the diagonal is called the **multiplicity** of the morphism; the diagonal block 0 indicates the $r \times r$ zero matrix, where $n = mk + r$. When $r = 0$, we obtain the unital morphisms. Injective morphisms $\mathcal{B}(\mathcal{H}) \to \mathcal{B}(\mathcal{L})$ have a similar form, except that we need to allow infinite multiplicity and U is interpreted as a unitary operator from $\mathcal{H} \oplus \mathcal{H} \oplus \cdots \oplus \mathcal{H} \oplus \mathcal{H}_0$ to $\mathcal{L}$. When $\mathcal{H}$ is infinite-dimensional, the algebra $\mathcal{B}(\mathcal{H})$ is not simple and the morphisms are much more complicated. However the algebra of compact operators is simple and the morphisms $\mathcal{K}(\mathcal{H}) \to \mathcal{K}(\mathcal{L})$ have essentially the same form as in the matrix case.

1.1.3

Morphisms $A \to B$ of finite direct sums of matrix algebras are described by their **Bratteli diagrams**. If

$$A = M_{k_1}(\mathbf{C}) \oplus \cdots \oplus M_{k_r}(\mathbf{C}) \quad \text{and} \quad B = M_{n_1}(\mathbf{C}) \oplus \cdots \oplus M_{n_s}(\mathbf{C})$$

then the Bratteli diagram consists of two rows of vertices, labeled by the k_i and the n_j respectively, together with m_{ij} edges connecting k_i and n_j for all i and j, where m_{ij} is the multiplicity of the composition

$$M_{k_i}(\mathbf{C}) \to A \to B \to M_{n_j}(\mathbf{C})$$

as described above. These integers are subject to the constraints

$$\sum_{i=1}^{r} k_i m_{ij} \leq n_j \qquad \text{for } 1 \leq j \leq s$$

Up to an inner automorphism of B (i.e., conjugation by a unitary element of B), the morphism represented by such a diagram takes $a = (a_1, \ldots, a_r)$ to $b = (b_1, \ldots, b_s)$, where $b_j \in M_{n_j}(\mathbf{C})$ is the diagonal block matrix with m_{1j}

copies of a_1, m_{2j} copies of a_2, and so forth down the diagonal. For example,

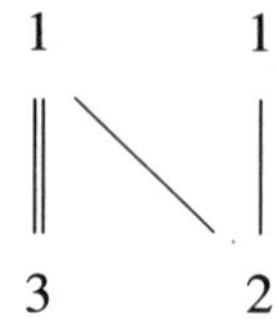

denotes the morphism $\mathbf{C} \oplus \mathbf{C} \to M_3(\mathbf{C}) \oplus M_2(\mathbf{C})$ given by

$$(\lambda, \mu) \to \left(\begin{pmatrix} \lambda & 0 & 0 \\ 0 & \lambda & 0 \\ 0 & 0 & 0 \end{pmatrix}, \begin{pmatrix} \lambda & 0 \\ 0 & \mu \end{pmatrix} \right)$$

1.2 UNITIZATION

Recall the following standard procedure for adjoining a unit to a complex algebra A:

$$A^+ = A \oplus \mathbf{C} \qquad \text{as a vector space}$$

$$(a, \lambda)(b, \mu) = (ab + \lambda b + \mu a, \lambda\mu) \qquad \text{for all } a, b \in A \text{ and } \lambda, \mu \in \mathbf{C}$$

Then A^+ is a unital algebra called the **unitization** of A. The unit element is $(0, 1)$ and $a \to (a, 0)$ is an embedding (an injective morphism) of A in A^+. If A has an involution then setting

$$(a, \lambda)^* = (a^*, \overline{\lambda}) \qquad \text{for all } a \in A, \lambda \in \mathbf{C}$$

extends it to an involution of A^+. A norm on A extends to a norm on A^+ via

$$\|(a, \lambda)\| = \sup\{\|ab + \lambda b\| : \|b\| \leq 1\} \qquad \text{for all } a \in A, \lambda \in \mathbf{C}$$

and if the norm on A satisfies the C^*-identity so does the extension:

$$\begin{aligned} \|(a, \lambda)^*(a, \lambda)\| &= \|(a^*a + \overline{\lambda}a + \lambda a^*, \overline{\lambda}\lambda)\| \\ &= \sup\{\|a^*ab + \overline{\lambda}ab + \lambda a^*b + \overline{\lambda}\lambda b\| : \|b\| \leq 1\} \\ &\geq \sup\{\|b^*a^*ab + \overline{\lambda}b^*ab + \lambda b^*a^*b + \overline{\lambda}\lambda b^*b\| : \|b\| \leq 1\} \end{aligned}$$

$$= \sup\{\|(ab + \lambda b)^*(ab + \lambda b)\| : \|b\| \le 1\}$$

$$= \sup\{\|ab + \lambda b\|^2 : \|b\| \le 1\}$$

$$= \|(a, \lambda)\|^2$$

We note also that the embedding $a \to (a, 0)$ is an isometry.

1.2.1

If A is already unital, with unit element 1, then $(a, \lambda) \to (a + \lambda 1, \lambda)$ is an isomorphism of A^+ with the direct sum $A \oplus \mathbf{C}$ as algebras. Note that the element $\lambda 1$ of A is customarily written simply as λ; this identifies $\mathbf{C}$ as a subalgebra of A.

1.2.2

For X locally compact and Hausdorff, the map

$$C_0(X)^+ \to C(X^+) : (f, \lambda) \to f + \lambda$$

where $X^+ = X \cup \{\infty\}$ is the one-point compactification and $(f + \lambda)(\infty) = \lambda$, is a $*$-isomorphism. Thus unitization can be viewed as a generalization of the one-point compactification construction from the commutative case to arbitrary C^*-algebras.

1.2.3

Any morphism $\varphi : A \to B$ of C^*-algebras extends uniquely to a unital morphism $\varphi^+ : A^+ \to B^+$, namely

$$\varphi^+(a, \lambda) = (\varphi(a), \lambda) \qquad \text{for all } a \in A, \lambda \in \mathbf{C}$$

In fact the map taking A to A^+ and φ to φ^+ is a **functor** from the category of C^*-algebras to the unital category. This means simply that $(\mathrm{id})^+ = \mathrm{id}$ and that composition works properly: $(\psi \circ \varphi)^+ = \psi^+ \circ \varphi^+$.

The functorial point of view helps explain the appearance of proper maps in 1.1.2, since the maps $X \to Y$ for which the extension $X^+ \to Y^+$ taking ∞ to ∞ is continuous are precisely the proper maps.

1.3 EXTENSIONS OF C^*-ALGEBRAS

Unitization provides an example of an extension of C^*-algebras. In general, an **extension** of C by A is a short exact sequence

$$0 \to A \xrightarrow{i} B \xrightarrow{\pi} C \to 0$$

of C^*-algebras and $*$-homomorphisms. In detail, this means that i is injective, that $i(A) = \ker \pi$ (so that $i(A)$ is an ideal of B) and that π is surjective. For example, the unitization extension is

$$0 \to A \xrightarrow{i} A^+ \xrightarrow{c} \mathbf{C} \to 0$$

with $i(a) = (a, 0)$ and $c(a, \lambda) = \lambda$. Extensions of C by A are **equivalent** if there is a commuting diagram

$$\begin{array}{ccccccccc} 0 & \longrightarrow & A & \longrightarrow & B_1 & \longrightarrow & C & \longrightarrow & 0 \\ & & {\scriptstyle =}\downarrow & & \downarrow & & \downarrow{\scriptstyle =} & & \\ 0 & \longrightarrow & A & \longrightarrow & B_2 & \longrightarrow & C & \longrightarrow & 0 \end{array}$$

Note that then $B_1 \to B_2$ is necessarily an isomorphism, so equivalence is an equivalence relation. This notion of equivalence (from homological algebra) is not the only possibility. For example, one might ask for an isomorphism $B_1 \to B_2$ that makes only the right-hand square commute. It turns out that the requirement appropriate to C^*-algebras is intermediate between these two; we will return to this later. The set of equivalence classes of extensions of C by A is denoted $\mathrm{Ext}(C, A)$.

1.3.1

Any extension $0 \to A \to B \to \mathbf{C} \to 0$ with $\mathbf{C}$ the complex numbers and B unital is equivalent to the unitization extension. For example, the unitization extension for $C_0(X)$ is equivalent to

$$0 \to C_0(X) \to C(X^+) \xrightarrow{\pi} \mathbf{C} \to 0$$

where π is evaluation at ∞.

Chapter Two

Spectrum and Order

2.1 SPECTRUM AND SPECTRAL RADIUS

For A unital the **spectrum** of $a \in A$ is

$$\sigma(a) = \sigma_A(a) = \{\lambda \in \mathbf{C} : a - \lambda \text{ is not invertible in } A\}$$

Note that for $a \in A$ we have $\sigma_{A^+}(a) = \sigma_A(a) \cup \{0\}$. When A is nonunital, we define $\sigma(a) = \sigma_{A^+}(a)$. The **spectral radius** is defined by

$$r(a) = \sup\{|\lambda| : \lambda \in \sigma(a)\}.$$

We recall that in any Banach algebra the spectrum is compact and nonempty, that $r(a) \leq \|a\|$ for all a, and that the spectral radius is given by the formula $r(a) = \lim \|a^n\|^{1/n}$.

2.1.1

If $\varphi : A \to B$ is unital and $B \neq 0$, then $\sigma(\varphi(a)) \subset \sigma(a)$ for all $a \in A$. In general, one has $\sigma(\varphi(a)) \subset \sigma(a) \cup \{0\}$.

2.1.2

An element u of a unital C^*-algebra A is **unitary** if $u^*u = 1 = uu^*$. Then

$$\|u\|^2 = \|u^*u\| = \|1\| = 1$$

if $A \neq 0$, and consequently $r(u) \leq \|u\| = 1$. Hence $\sigma(u) \subset \{\lambda : |\lambda| \leq 1\}$. It follows that $\sigma(u^{-1}) \subset \{\lambda : |\lambda| \geq 1\}$. But $u^{-1} = u^*$ is also unitary, and therefore $\sigma(u)$ is contained in the unit circle.

2.1.3

An element h of a C^*-algebra A is **self-adjoint** or **hermitian** if $h^* = h$. Any $a \in A$ is uniquely of the form $a = h + ik$ with h and k self-adjoint.

Theorem

If h is self-adjoint then $r(h) = \|h\|$.

Proof

$\|h\|^2 = \|h^*h\| = \|h^2\|$ and by induction $\|h\|^{2^n} = \|h^{2^n}\|$ for all n. Take 2^nth roots, let $n \to \infty$ and use the spectral radius formula in 2.1 to get the result.

2.1.4 *Corollary*

An algebra with involution admits at most one norm making it a C^*-algebra.

Proof

For any element a one has $\|a\|^2 = \|a^*a\| = r(a^*a)$, since a^*a is self-adjoint.

2.1.5 *Corollary*

Any morphism $\varphi\colon A \to B$ of C^*-algebras is norm-decreasing, hence continuous.

Proof

It follows from the C^*-identity that we only need to show $\|\varphi(h)\| \le \|h\|$ for h self-adjoint. By 2.1.1 we have $\sigma(\varphi(h)) \subset \sigma(h) \cup \{0\}$, and consequently $r(\varphi(h)) \le r(h)$. The theorem now gives

$$\|\varphi(h)\| = r(\varphi(h)) \le r(h) = \|h\|$$

2.1.6 *Corollary*

The kernel of any morphism of C^*-algebras is a closed *-ideal.

2.1.7 *Theorem*

If h is self-adjoint then $\sigma(h)$ is real.

Proof

The series $1+(ih)+\frac{1}{2!}(ih)^2+\cdots$ converges and defines an element e^{ih} of A (or of A^+, if A is nonunital). Since involution is continuous, we have $(e^{ih})^* = e^{-ih}$. In any Banach algebra the identity $e^a e^b = e^{a+b}$ is valid whenever a and b commute. Consequently e^{ih} is unitary, and by 2.1.2 its spectrum is contained in the unit circle. But, if $\lambda \in \sigma(h)$, then for some a

$$e^{ih} - e^{i\lambda} = e^{i\lambda}(e^{i(h-\lambda)} - 1) = e^{i\lambda}(h - \lambda)a$$

Hence $e^{ih} - e^{i\lambda}$ is not invertible (because $h - \lambda$ is not invertible and commutes with a); therefore $e^{i\lambda}$ is on the unit circle, and so λ is real.

2.1.8 Theorem

If B is a unital C^*-algebra and A is a unital C^*-subalgebra of B (same unit) then $\sigma_A(a) = \sigma_B(a)$ for all $a \in A$.

Proof

By general Banach algebra theory one has $\sigma_A(a) \supset \sigma_B(a)$, with equality whenever $\sigma_B(a)$ has no holes. For the reverse inclusion it is sufficient to show that $a \in A$ and a invertible in B together imply that a is invertible in A. But, if a is invertible in B, so is aa^*. Then aa^* is invertible in A, since $\sigma_B(aa^*)$ is real by 2.1.7 and thus has no holes. Therefore a has a right inverse in A. But this right inverse must equal the inverse of a in B and so a is invertible in A.

2.1.9 Corollary

If B is a C^*-algebra and A is a C^*-subalgebra of B, then $\sigma_A(a) \cup \{0\} = \sigma_B(a) \cup \{0\}$ for all $a \in A$.

Proof

Apply 2.1.8 to B^+ and the C^*-subalgebra generated by A and the unit of B^+.

2.2 THE GELFAND-NAIMARK THEOREM (COMMUTATIVE ALGEBRAS)

Let A be a commutative unital Banach algebra, and let $\hat{A} \subset A^*$ be the set of nonzero bounded multiplicative linear functionals on A. We recall that $\hat{A}$ is compact and Hausdorff in the relative weak*-topology and that the **Gelfand**

transform $a \to \hat{a}: A \to C(\hat{A})$, defined by $\hat{a}(\varphi) = \varphi(a)$ for all $a \in A$ and $\varphi \in \hat{A}$, is a norm-decreasing (in fact $\|\hat{a}\|_\infty = r(a)$) homomorphism of complex algebras.

2.2.1

Observe that if A is a commutative unital C^*-algebra, then any $\varphi \in \hat{A}$ is in fact a unital morphism $A \to \mathbf{C}$ of C^*-algebras. It is unital because it is multiplicative, and it follows that $\varphi(a) \in \sigma(a)$ for all $a \in A$ (because $\varphi(a - \varphi(a)) = 0$). Thus $\varphi(h)$ is real if h is self-adjoint by 2.1.7. If $a = h + ik$ with h and k self-adjoint as in 2.1.3, then

$$\varphi(a^*) = \varphi(h - ik) = \varphi(h) - i\varphi(k) = \overline{\varphi(h) + i\varphi(k)} = \overline{\varphi(a)}$$

as asserted.

2.2.2 Theorem [GeN43]

For any commutative unital C^*-algebra A, the Gelfand transform $A \to C(\hat{A})$ is an isometric $*$-isomorphism.

Proof

The Gelfand transform is $*$-preserving by 2.2.1. It is isometric because

$$\|\hat{a}\|_\infty^2 = \|\hat{a}^*\hat{a}\|_\infty = \|\widehat{a^*a}\|_\infty = r(a^*a) = \|a^*a\| = \|a\|^2$$

for all $a \in A$, using Theorem 2.1.3. It is onto by the Stone-Weierstrass theorem, since the range is a uniformly closed self-adjoint subalgebra of $C(\hat{A})$ that contains the constants and separates the points of $\hat{A}$.

2.2.3 Corollary

Any injective morphism $\varphi: A \to B$ of C^*-algebras is an isometry.

Proof

It is enough to show $\|\varphi(a)\| = \|a\|$ for a self-adjoint. We can assume that A, B, and φ are unital by passing to $\varphi^+: A^+ \to B^+$ as in 1.2.3. We can also assume that A and B are commutative by dropping down to $\varphi: C^*(a) \to C^*(\varphi(a))$, where $C^*(a)$ denotes the C^*-subalgebra generated by a and 1. Thus by the theorem we can suppose that we have a unital injection $\varphi: C(X) \to C(Y)$

for some compact Hausdorff spaces X and Y. It follows (see 1.1.2) that there is a continuous surjection $\alpha : Y \to X$ such that

$$\varphi(f) = f \circ \alpha \qquad \text{for all } f \in C(X)$$

Then $\|\varphi(f)\|_\infty = \|f\|_\infty$ because both functions have the same range.

2.2.4

Theorem 2.2.2 is known as the Gelfand-Naimark theorem. It can be extended to arbitrary commutative C^*-algebras A by applying 2.2.2 to the unitization A^+. The result is that the Gelfand transform is an isometric $*$-isomorphism from A onto $C_0(\hat{A})$, where $\hat{A}$ (still the space of nonzero multiplicative linear functionals on A) is in general only locally compact.

2.2.5

If A is a commutative unital C^*-algebra, and if A is singly generated, say by $a \in A$, then $A \cong C(\sigma(a))$. In other words, in this situation $\hat{A}$ is homeomorphic to $\sigma(a)$. In fact $\hat{a} : \varphi \to \varphi(a)$ is such a homeomorphism. It is a basic property of the Gelfand transform that for any element a of a commutative unital Banach algebra, the range of $\hat{a}$ is $\sigma(a)$ [Dou72, 2.36]; when a generates, $\hat{a}$ is evidently one-to-one. Note that this isomorphism $A \cong C(\sigma(a))$ carries a multiplicative linear functional φ on A to evaluation at $\varphi(a)$ on $C(\sigma(a))$. The inverse map $C(\sigma(a)) \to A$ is usually written $f \to f(a)$ and is called the **functional calculus** of a.

2.2.6

Gelfand and Naimark [GeN43] also proved a representation theorem of fundamental importance that applies to general (i.e., not necessarily commutative) C^*-algebras. This result, also known as the Gelfand-Naimark theorem, states that any C^*-algebra can be realized as a C^*-subalgebra of some $\mathcal{B}(\mathcal{H})$. It will be treated in Chapter 5.

2.3 POSITIVITY

An element $a \in A$ of a C^*-algebra is **positive** if it is self-adjoint and $\sigma(a)$ is contained in $[0, +\infty)$. We write $a \geq 0$, and we let

$$A_+ = \{a \in A : a \geq 0\}$$

This should not be confused with the notation A^+ for the unitization. We will see below that A_+ is a cone, the **positive cone** of A. As is customary, $a \leq b$ means $b - a \in A_+$. For $f \in C(X)$ we see that $f \geq 0$ has the usual meaning because $\sigma(f) = \{f(x) : x \in X\}$. In $\mathcal{B}(\mathcal{H})$ we have $T \geq 0$ if and only if T is self-adjoint and $\langle T\xi, \xi\rangle \geq 0$ for all $\xi \in \mathcal{H}$ (because for T self-adjoint the convex hull of $\sigma(T)$ is equal to the numerical range of T).

2.3.1

Note in particular that $a \leq \|a\|$ for any self-adjoint element a. In other words, $\|a\| - a \geq 0$, which follows from

$$\sigma(\|a\| - a) = \{\|a\| - \lambda : \lambda \in \sigma(a)\}$$

with the help of 2.1.3 and 2.1.7.

2.3.2

For any self-adjoint $a \in A$,

$$a \in A_+ \Longleftrightarrow a = b^2 \qquad \text{for some } b \in A_+$$

For the forward implication, drop down to the C^*-subalgebra $C^*(a)$ generated by a, observe that a remains positive there by 2.1.9, and apply the Gelfand-Naimark theorem 2.2.2 (and 2.2.4). For the reverse implication proceed similarly with $C^*(b)$.

2.3.3

Positive square roots are unique. Suppose that $a = b^2 = c^2$ with a and b as above and $c \in A_+$. Then $ca = ac$, and since $b \in C^*(a)$, it follows that $cb = bc$. Thus the C^*-subalgebra $C^*(b, c)$ generated by b and c is commutative, so the Gelfand-Naimark theorem applies to reduce the question to the case of functions, where it is obviously true.

2.3.4 Theorem

The set A_+ is a cone; in particular $\leq$ is a partial ordering of A.

Proof

Clearly $\lambda a \in A_+$ whenever $a \in A_+$ and $\lambda \geq 0$. To show that $a, b \in A_+$ implies $a + b \in A_+$, it is convenient to use the following characterization of positivity: For any self-adjoint contraction $a \in A$,

$$a \in A_+ \iff \|1 - a\| \leq 1$$

This is true for functions and thus in any C^*-algebra via the Gelfand-Naimark theorem. Now any $a, b \in A_+$ can be replaced by suitable multiples ta, tb if necessary to make both contractions; then one has

$$\|1 - \tfrac{1}{2}(a + b)\| = \tfrac{1}{2}\|(1 - a) + (1 - b)\| \leq 1$$

so that $\frac{1}{2}(a + b) \in A_+$ and $a + b \in A_+$ by the above characterization of positivity. Finally, $A_+ \cap (-A_+) = \{0\}$ since, if both a and $-a$ are in A_+, then $\sigma(a) = \{0\}$, and so $a = 0$ by 2.1.3.

2.3.5 Corollary

$0 \leq a \leq b \Longrightarrow \|a\| \leq \|b\|$.

Proof

We have $b \leq \|b\|$ by 2.3.1, and so $a \leq \|b\|$ by the theorem. Hence $r(a) \leq \|b\|$, and therefore $\|a\| \leq \|b\|$ by 2.1.3.

2.4 POSITIVITY OF a^*a

To proceed, we need to know that $a^*a \in A_+$ for all $a \in A$. One reason for this is that it is true for operators on a Hilbert space, and we want to represent C^*-algebras as algebras of operators. Since a^*a is self-adjoint, this condition is equivalent to the invertibility of $1 + a^*a$ for all a. The latter was taken as an axiom in the original paper of Gelfand and Naimark [GeN43] but was soon found to be a consequence of the other axioms (Kaplansky's monograph [Kap70] contains interesting historical material about the axiomatization of C^*-algebras). Here is the proof.

Let $b = a^*a$, and decompose $b = b^+ - b^-$ with $b^\pm \in A_+$ and $b^+b^- = 0$ using the Gelfand-Naimark theorem. If $c = ab^-$, then

$$-c^*c = -b^-a^*ab^- = -b^-(b^+ - b^-)b^- = (b^-)^3 \in A_+$$

We want to show that this forces $c = 0$, for then

$$0 = a^*c = a^*ab^- = (b^+ - b^-)b^- = -(b^-)^2$$

$b^- = 0$ and $a^*a = b^+ \in A_+$. Write $c = h + ik$ with h, k self-adjoint. Then

$$c^*c + cc^* = 2h^2 + 2k^2$$

and since $-c^*c \in A_+$, we have $cc^* \in A_+$ by 2.3.4. However,

$$\sigma(c^*c) \cup \{0\} = \sigma(cc^*) \cup \{0\}$$

(see [KaR83, 3.2.8]), so $c^*c \in A_+$. Hence $\|c\|^2 = \|c^*c\| = r(c^*c) = 0$, and so $c = 0$.

2.4.1 *Theorem*

$a \leq b \Longrightarrow x^*ax \leq x^*bx$ *for all* a, b, x.

Proof

We have $b - a = h^2$ for some $h \in A_+$ by 2.3.2. Then

$$x^*bx - x^*ax = x^*h^2x = (hx)^*(hx) \in A_+$$

by what was shown above.

2.4.2 *Corollary*

The inequality $x^*a^*ax \leq \|a\|^2x^*x$ *holds for all* a, x.

2.4.3 *Corollary*

If $0 \leq a \leq b$ *and* a *is invertible, then* b *is invertible and* $0 \leq b^{-1} \leq a^{-1}$.

Proof

The assertion is true when $b = 1$ by the Gelfand-Naimark theorem (2.2.2). For the general case, observe first that for any $c \in A^+$,

$$c \text{ invertible } \Longleftrightarrow c \geq \varepsilon \text{ for some real } \varepsilon > 0$$

From this and 2.3.4 it follows that b is invertible. Consequently so is its (unique, positive) square root $b^{1/2}$. We now have

$$a \leq b \Longrightarrow b^{-1/2}\, ab^{-1/2} \leq 1 \qquad \text{by the theorem}$$

$$\Longrightarrow 1 \le (b^{-1/2}\, ab^{-1/2})^{-1} = b^{1/2}a^{-1}b^{1/2}$$

$$\Longrightarrow b^{-1} \le a^{-1} \qquad \text{by the theorem again.}$$

2.4.4

There has been some interest in functions $f : \mathbf{R} \to \mathbf{R}$ that are "operator monotone" in the sense that $f(a) \le f(b)$ whenever a, b are self-adjoint and $a \le b$. Löwner [Löw34] and Bendat and Sherman [BeS55] have obtained characterizations in the matrix and operator cases.

Square rooting is operator monotone but, surprisingly, squaring is not. The matrices

$$p = \begin{pmatrix} 1 & 0 \\ 0 & 0 \end{pmatrix} \quad \text{and} \quad q = \frac{1}{2}\begin{pmatrix} 1 & 1 \\ 1 & 1 \end{pmatrix}$$

satisfy $0 \le p \le p + q$ but not $p^2 \le (p+q)^2$.

2.5 APPROXIMATE UNITS

An **approximate unit** for a C^*-algebra A is a net $\{u_\lambda : \lambda \in \Lambda\}$ of positive contractions in A such that

i. $\lambda \le \mu \Longrightarrow u_\lambda \le u_\mu$ for all $\lambda, \mu \in \Lambda$.
ii. $\|a - au_\lambda\|, \|a - u_\lambda a\| \to 0$ for all $a \in A$.

The notation (u_λ) is also in use. A **sequential approximate unit** is an approximate unit indexed by the positive integers.

2.5.1

If $\{e_1, e_2, \ldots\}$ is an orthonormal basis for a Hilbert space $\mathcal{H}$ and P_n is the orthogonal projection on the span of $\{e_1, e_2, \ldots, e_n\}$, then $\{P_n\}$ is an approximate unit for the algebra $\mathcal{K}(\mathcal{H})$ of compact operators on $\mathcal{H}$—but not for $\mathcal{B}(\mathcal{H})$.

For another example, let X be σ-compact and Hausdorff, so that one has $X = \cup_{n=1}^{\infty} O_n$ with O_n open, $\overline{O}_n$ compact and $\overline{O}_n \subset O_{n+1}$ for all n. Choose $u_n \in C_0(X)$ with $u_n = 1$ on $\overline{O}_n$ and $u_n = 0$ off O_{n+1}. Then $\{u_n\}$ is an approximate unit for $C_0(X)$. In general,

$$\{u \in C_0(X) : 0 \le u \le 1\}$$

is an approximate unit for $C_0(X)$. Something like this is true in any C^*-algebra:

2.5.2 *Theorem*

For any C^*-algebra A, the set

$$\Lambda = A_+ \cap \{a \in A : \|a\| < 1\}$$

is upward-directed and an approximate unit.

Proof

To see that Λ is directed, we observe that A_+ is directed (by 2.3.4) and order isomorphic to Λ. In fact $a \to 1 - (1+a)^{-1}$ defines an order isomorphism by 2.4.3. Note here that the element $1 - (1+a)^{-1}$, computed in the unitization of A, actually belongs to A.

To prove that Λ is an approximate unit, it will be enough to show that $u_\lambda a \to a$ for all $a \in \Lambda$. Given $\varepsilon > 0$ and $a \in \Lambda$, let

$$u_\varepsilon = 1 - (1 + \varepsilon^{-2}a)^{-1} = \varepsilon^{-2}a(1 + \varepsilon^{-2}a)^{-1}$$

and consider any $u \in \Lambda$, $u \geq u_\varepsilon$. We have

$$(1-u)^2 \leq 1 - u \leq 1 - u_\varepsilon = (1 + \varepsilon^{-2}a)^{-1}$$

and so by 2.4.1,

$$a(1-u)^2 a \leq a(1 + \varepsilon^{-2}a)^{-1}a \leq \varepsilon^2 a$$

Hence $\|a - ua\| < \varepsilon$ by 2.3.5.

2.5.3

If A is separable, then it has a sequential approximate unit. Take $\{a_1, a_2, \ldots\}$ dense in A, and choose elements $u_1, u_2, \ldots$ of the set Λ of 2.5.2 recursively such that

$$u_1 \leq u_2 \leq \cdots \quad \text{and} \quad \|a_k - u_n a_k\| < \frac{1}{n} \qquad \text{for } 1 \leq k \leq n$$

Then (u_n) is an approximate unit for A.

2.5.4 Theorem

Let I be a closed two-sided ideal of a C^*-algebra A. Then I is a $*$-ideal and, with $(a + I)^* = a^* + I$ and the quotient Banach algebra structure, A/I is a C^*-algebra.

Proof

Let $b \in I$, and observe that $C^*(b^*b) \subset I$. By 2.5.2 we can choose an approximate unit $\{u_\lambda\}$ for $C^*(b^*b)$. Since

$$\|b(1 - u_\lambda)\|^2 = \|(1 - u_\lambda)b^*b(1 - u_\lambda)\| \leq \|b^*b(1 - u_\lambda)\|$$

we have $bu_\lambda \to b$ in norm. Thus $u_\lambda b^* \to b^*$ in norm, and so $b^* \in I$. The C^*-identity for the quotient norm follows from the formula

$$\|a + I\| = \lim \|a - au_\lambda\|$$

valid for any $a \in A$ and approximate unit (u_λ) for the C^*-algebra I. In fact

$$\begin{aligned} \|a + I\|^2 &= \lim \|a(1 - u_\lambda)\|^2 \\ &= \lim \|(1 - u_\lambda)a^*a(1 - u_\lambda)\| \\ &\leq \lim \|a^*a(1 - u_\lambda)\| \\ &= \|a^*a + I\| = \|(a + I)^*(a + I)\| \end{aligned}$$

To prove the formula, note that for any $b \in I$,

$$\begin{aligned} \|a - au_\lambda\| &\leq \|(a + b) - (a + b)u_\lambda\| + \|b - bu_\lambda\| \\ &\leq \|a + b\| + \|b - bu_\lambda\| \end{aligned}$$

so that $\limsup \|a - au_\lambda\| \leq \|a + I\|$. But $\|a + I\| \leq \liminf \|a - au_\lambda\|$, since $au_\lambda \in I$.

2.5.5 Corollary

Any morphism of C^*-algebras has closed range.

Proof

Any morphism $\varphi : A \to B$ of C^*-algebras induces an injective morphism $A/I \to B$, where I is the kernel of φ. Since I is closed by 2.1.6, the assertion follows from the theorem and 2.2.3.

2.5.6 *Corollary*

For any C^*-subalgebra B and closed two-sided ideal I of a C^*-algebra A, $B + I$ is a C^*-subalgebra of A.

Proof

Consider the restriction of the quotient map $A \to A/I$ to B. Its range is $(B + I)/I$ which is closed by 2.5.5. It follows that $B + I$ is closed in A, as it is the preimage in A of $(B + I)/I$.

Chapter Three

Examples and Constructions

3.1 INTRODUCTION

We have already introduced in 1.1.1 the commutative C^*-algebras $C_0(X)$, where X is a locally compact Hausdorff space, and we have seen in 2.2.2 and 2.2.4 that conversely any commutative C^*-algebra arises in this way. We have also indicated that the matrix algebras $M_n(\mathbf{C})$ and their infinite-dimensional analogues, the algebras $\mathcal{B}(\mathcal{H})$ of bounded linear transformations on a Hilbert space $\mathcal{H}$, are basic examples of C^*-algebras. We will see later that in fact any C^*-algebra is isometrically $*$-isomorphic to a C^*-subalgebra of some $\mathcal{B}(\mathcal{H})$.

3.2 ALGEBRAS WITH COEFFICIENTS

The examples in 3.1 can be generalized by allowing "coefficients" from an arbitrary C^*-algebra A. Let X be locally compact and Hausdorff, and let $C_0(X, A)$ be the algebra of continuous functions $f : X \to A$ that vanish at infinity (i.e., $\{x \in X : \|f(x)\| \geq \varepsilon\}$ is compact for all $\varepsilon > 0$). With

$$f^*(x) = (f(x))^* \quad \text{and} \quad \|f\| = \max\{\|f(x)\| : x \in X\}$$

this is a C^*-algebra. The algebra $M_n(A)$ of $n \times n$ matrices over A is also a C^*-algebra. The involution is given by $(a^*)_{ij} = (a_{ji})^*$, and to define a C^*-norm, we observe that if A is realized as operators on a Hilbert space $\mathcal{H}$, as mentioned in 3.1, then $M_n(A)$ can be regarded as operators on the n-fold direct sum $\mathcal{H}^n$ of copies of $\mathcal{H}$ or, in other words, as a C^*-subalgebra of $\mathcal{B}(\mathcal{H}^n)$. Notice also that $M_n(C_0(X, A))$ can be identified with $C_0(X, M_n(A))$.

3.3 DIRECT SUMS

The direct sum (or product) $A = A_1 \oplus \cdots \oplus A_n$ of C^*-algebras, with the pointwise algebraic operations and norm

$$\|(a_1, \ldots, a_n)\| = \max\{\|a_1\|, \ldots, \|a_n\|\}$$

is a C^*-algebra. For infinite families $\{A_i\}_{i \in I}$ of C^*-algebras, there is a distinction between direct sums and direct products. The **direct product** or **ℓ^∞-direct sum** consists of elements a of the algebraic direct product such that $i \to \|a_i\|$ is bounded. For the **direct sum** or **c_0-direct sum**, we require that $i \to \|a_i\|$ vanish at infinity on the discrete set I. In either case the norm is the supremum of this function. When all $A_i = \mathbf{C}$, we get the classical algebras $\ell^\infty(I)$ and $c_0(I)$. When all $A_i = A$, a fixed C^*-algebra, we get $C_b(I, A)$ and $C_0(I, A)$ where C_b denotes the algebra of bounded continuous functions. We note also that any finite-dimensional C^*-algebra is $*$-isomorphic to a finite direct sum of matrix algebras $M_n(\mathbf{C})$ for suitable values of n. To see this, it is helpful to have some knowledge of the representation theory of C^*-algebras; we will return to it in Chapter 5.

3.4 ALGEBRAS FROM BUNDLES

A complex vector bundle E is, roughly speaking, a parametrized family of complex vector spaces. In suitable circumstances there will be a C^*-algebra bearing the same relationship to E that $M_n(\mathbf{C})$ does to $\mathbf{C}^n$. We recall some of the details [Ati67; Hus66].

Let X be a topological space (usually compact Hausdorff). A **product bundle** over X is a product $X \times \mathbf{C}^n$ together with the projection $X \times \mathbf{C}^n \to X$ on the first coordinate. A **complex vector bundle** over X is a topological space E and a continuous surjection $p : E \to X$, together with the structure of a complex vector space on each **fiber** $E_x = p^{-1}(x)$, which is **locally trivial**. This means that each $x \in X$ has a neighborhood U such that the **restriction** $E|U = p^{-1}(U)$ is isomorphic to a product bundle $U \times \mathbf{C}^n$. Here the isomorphism is to be understood in the sense appropriate to vector bundles: a **morphism** $E \to F$ of vector bundles over X is a continuous map that takes E_x linearly to F_x for all $x \in X$.

Now observe that the space $\mathrm{Hom}(E, F)$ of all morphisms $E \to F$ is a complex vector space and, with composition as multiplication, a complex algebra when $E = F$. To give this the structure of a C^*-algebra, we first define the **direct sum** or **Whitney sum** of vector bundles $p : E \to X$ and

$q : F \to X$ as the **pullback**

$$E \oplus F = \{(e, f) \in E \times F : p(e) = q(f)\}$$

This is a vector bundle over X with fibers $E_x \oplus F_x$. A **hermitian metric** on E is then a continuous map $E \oplus E \to \mathbf{C}$ such that the restriction to each $E_x \oplus E_x$ is an inner product on E_x, and a **hermitian vector bundle** is a complex vector bundle together with a hermitian metric. If E and F are hermitian vector bundles, $T \in \mathrm{Hom}(E, F)$ and $T_x : E_x \to F_x$ is the restriction, then T_x^* and $\|T_x\|$ make sense, and we can define T^* by $(T^*)_x = T_x^*$ and $\|T\| = \max \|T_x\|$, the latter when X is compact. Consequently, if E is a hermitian vector bundle over a compact space X, then $\mathrm{Hom}(E, E)$ is naturally a C^*-algebra. When E is the product bundle $X \times \mathbf{C}^n$, it is easy to see that $\mathrm{Hom}(E, E)$ is $C(X, M_n(\mathbf{C}))$. We will see later that in general one gets new C^*-algebras (see 6.3 below).

3.5 *ALGEBRAS FROM CONTINUOUS FIELDS*

The construction of the previous section has a useful generalization. A **continuous field of C^*-algebras** over a topological space X is an assignment $x \to A_x$ of C^*-algebras to the points of X, together with a $*$-subalgebra Γ of the cartesian product $\prod A_x$, such that

i. $\{a(x) : a \in \Gamma\}$ is dense in A_x for all $x \in X$,

ii. $x \to \|a(x)\|$ is continuous for all $a \in \Gamma$,

iii. Γ is locally uniformly closed.

For example, $x \to \mathrm{Hom}(E_x, E_x)$, E a hermitian vector bundle, is a continuous field of C^*-algebras with $\Gamma = \mathrm{Hom}(E, E)$.

To any such field over a locally compact space, one associates the C^*-algebra of all $a \in \Gamma$ such that the function in condition ii vanishes at infinity. Some examples are given in Chapter 8. See [Dix77, ch.10] for an extensive development of this topic.

3.6 *GROUP C^*-ALGEBRAS*

In this section we will look at how C^*-algebras arise and are used in the study of locally compact groups.

3.6.1

For a finite group G and a ring R, the **group ring** RG is defined to be the set of all functions $G \to R$ with pointwise addition and the convolution product

$$(xy)(g) = \sum_h x(h)y(h^{-1}g) = \sum_{g=hk} x(h)y(k)$$

for all $x, y \in RG$ and $g \in G$. For example, if R is unital and we denote by δ_g the characteristic function of $\{g\}$, then $\delta_g \delta_h = \delta_{gh}$ for all $g, h \in G$. Since the δ_g span RG, it is easy to see that this special case implies the general formula. Alternatively, RG can be viewed as consisting of all formal sums $\sum_{g \in G} r_g u_g$ with $r_g \in R$, these objects being multiplied by agreeing that r_g commutes with u_h and that $u_g u_h = u_{gh}$ for all $g, h \in G$.

An involution on R may be used to define an involution on RG. For example, if $R = \mathbf{C}$, then

$$x^*(g) = \overline{x(g^{-1})} \qquad \text{for all } x \in \mathbf{C}G$$

gives an involution on $\mathbf{C}G$. In particular, $\delta_g^* = \delta_{g^{-1}}$, so δ_g is unitary for all $g \in G$.

We want to show that $\mathbf{C}G$ admits a norm that makes it a C^*-algebra. The idea is to represent $\mathbf{C}G$ as bounded operators on a Hilbert space and then to use the operator norm. But $\mathbf{C}G$ is itself a Hilbert space with inner product

$$\langle x, y \rangle = \sum_g x(g)\overline{y(g)} \qquad \text{for all } x, y \in \mathbf{C}G$$

and the **left regular representation** $\lambda : \mathbf{C}G \to \mathcal{B}(\mathbf{C}G)$, defined by

$$\lambda(x)y = xy \qquad \text{for all } x, y \in \mathbf{C}G$$

is an injective $*$-homomorphism. Of course in this case $\mathbf{C}G$ is finite-dimensional and hence is a finite direct sum of matrix algebras $M_n(\mathbf{C})$. But, as we will see later, the construction generalizes to give a C^*-algebra $C^*(G)$, called the **group C^*-algebra** of G, for any locally compact group G.

3.6.2

By a **unitary representation** of a locally compact group G is meant a homomorphism $g \to U_g$ of G into the unitary group $\mathcal{U}(\mathcal{H})$ of a Hilbert space $\mathcal{H}$ such that $g \to U_g\xi$ is continuous for any $\xi \in \mathcal{H}$. The group C^*-algebra

$C^*(G)$ should be thought of as encoding all information about the unitary representations of G. In more precise terms this means that $C^*(G)$ has the universal property that any unitary representation $U : G \to \mathcal{U}(\mathcal{H})$ extends to a $*$-homomorphism $\pi : C^*(G) \to \mathcal{B}(\mathcal{H})$. This can be taken literally when G is discrete, since in this case $C^*(G)$ contains a copy of G in its unitary group. For example, when G is finite, the extension is given by $\pi(\sum \lambda_g u_g) = \sum \lambda_g U_g$. In general, one still has a bijection between unitary representations of G and (nondegenerate) representations of $C^*(G)$.

When G is abelian, so is $C^*(G)$, and thus it is isomorphic to some $C_0(X)$ by the Gelfand-Naimark theorem 2.2.2. Here the space X of nonzero multiplicative linear functionals on $C^*(G)$ may be identified with the space of homomorphisms from G to the circle group S^1 (the unitary group of $\mathbf{C}$). With respect to pointwise multiplication these homomorphisms themselves form a locally compact abelian group, denoted $\hat{G}$ and called the **dual group** of G, and so one has $C^*(G) \cong C_0(\hat{G})$.

3.6.3

For general locally compact groups G, there are two possibilities for the group C^*-algebra, the (full) algebra $C^*(G)$ and a quotient of it, the **reduced** algebra $C^*_r(G)$. These need not be the same. Suppose that G is infinite and discrete. Proceeding as in 3.6.1, but starting with the finitely nonzero functions $G \to \mathbf{C}$, the formulas there make sense and lead to the reduced algebra. In some cases this algebra does not capture all information about the unitary representations of G. If F_2 is the free group on two generators, then $C^*_r(F_2)$ is known to be simple [Pow75]. Since it is infinite-dimensional, it follows that it cannot have any finite-dimensional representations. But F_2 clearly has finite-dimensional unitary representations and hence so does $C^*(F_2)$. Thus $C^*_r(F_2)$ is a proper quotient of $C^*(F_2)$.

In his proof Powers makes use of the linear functional τ on $C^*_r(F_2)$ defined by

$$\tau(A) = \langle A\delta_e, \delta_e\rangle \qquad \text{for all } A \in C^*_r(F_2),$$

where $e \in G$ is the identity element. It is easily seen to be faithful (i.e., nonvanishing on the positive cone) and a trace ($\tau(AB) = \tau(BA)$ for all A and B and $\tau(I) = 1$). Powers shows that for any self-adjoint $A \in C^*_r(F_2)$ and $\varepsilon > 0$ there are group elements $g_1, \ldots, g_n$ and positive scalars $t_1, \ldots, t_n$ such that

$$\sum t_i = 1 \quad \text{and} \quad \left\| \tau(A)I - \sum t_i \lambda(u_{g_i}) A \lambda(u_{g_i^{-1}}) \right\| < \varepsilon$$

Now, if $\mathcal{I}$ is a nonzero ideal, then it contains a positive element A with $\tau(A) = 1$; applying the result to this A shows that $\mathcal{I}$ contains an invertible element and so is not proper. It also follows that the trace is unique: If τ' is any trace, then easily

$$\tau(A) - \tau'(A) = \tau'\left(\tau(A)I - \sum t_i \lambda(u_{g_i}) A \lambda(u_{g_i^{-1}})\right)$$

for all A, t_i, and g_i, and then the quoted result implies that $\tau = \tau'$.

We will return to these topics in Chapter 8.

3.7 TENSOR PRODUCTS

Given vector spaces V and W over a field F, it is not difficult to see that there exist a vector space X and a bilinear map $\varphi : V \times W \to X$ such that any bilinear map $\psi : V \times W \to Z$ factors uniquely through φ:

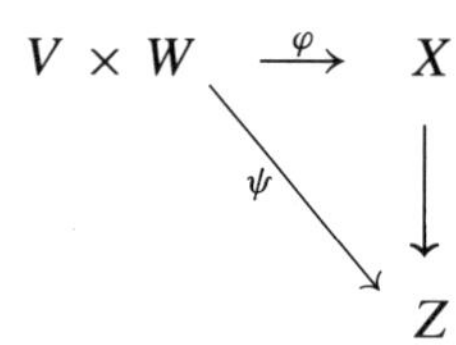

From the uniqueness of the factorization, it follows that the pair (X, φ) is unique up to isomorphism and that the range of φ spans X. The space X is called the **tensor product** of V and W, written $V \otimes W$. For $(v, w) \in V \times W$ the element $\varphi(v, w) \in V \otimes W$ is denoted by $v \otimes w$. As just mentioned, these **elementary tensors** span $V \otimes W$.

3.7.1

The notation V^* will be used for the dual space of a vector space V.

Lemma

To each pair $(f, g) \in V^* \times W^*$ corresponds a unique $h \in (V \otimes W)^*$ such that

$$h(v \otimes w) = f(v)g(w) \qquad \textit{for all } v \in V,\ w \in W$$

This results immediately from the fact that $(v, w) \to f(v)\,g(w)$ is a bilinear map $V \times W \to F$. The notation $f \otimes g$ is used to denote the functional h.

3.7.2

The representation of an element of a tensor product as a sum of elementary tensors is highly nonunique. However:

Lemma

Any element $u \in V \otimes W$ can be written $u = \sum v_i \otimes w_i$ so that the v_i are linearly independent; the w_i are then uniquely determined by the v_i.

Proof

As remarked just above 3.7.1 any $u \in V \otimes W$ can be written as a finite sum $u = \sum v_i \otimes w_i$. If, for example, $v_1 = \sum_{i\geq 2} \lambda_i v_i$, then

$$v_1 \otimes w_1 = \sum_{i\geq 2} v_i \otimes \lambda_i w_1 \qquad \text{and so}$$

$$u = \sum_{i\geq 2} v_i \otimes (w_i + \lambda_i w_1)$$

If $\sum v_i \otimes w_i = 0$ and the v_i are linearly independent, we choose $f_j \in V^*$ such that $f_j(v_i) = \delta_{ij}$. Then for all $g \in W^*$, we have

$$0 = (f_j \otimes g)\left(\sum v_i \otimes w_i\right) = g(w_j)$$

by 3.7.1 and therefore all $w_j = 0$.

3.7.3 Corollary

i. *For inner product spaces V and W, there is a unique inner product on $V \otimes W$ such that*

$$\langle v \otimes w, v' \otimes w' \rangle = \langle v, v' \rangle \langle w, w' \rangle \qquad \textit{for all } v, v' \in V \textit{ and } w, w' \in W$$

ii. *For algebras A and B, there is a (unique) product on $A \otimes B$ such that*

$$(a \otimes b)(a' \otimes b') = aa' \otimes bb' \qquad \textit{for all } a, a' \in A \textit{ and } b, b' \in B$$

iii. *If A and B have involutions, there is a unique involution on $A \otimes B$ such that*

$$(a \otimes b)^* = a^* \otimes b^* \qquad \textit{for all } a \in A \textit{ and } b \in B$$

Proof

Here is the proof of i; the others are similar. From bilinearity and the given condition, it is clear that there is only one possibility for the inner product. The only question is whether it is well-defined:

$$\sum_{i=1}^{n} v_i \otimes w_i = 0 \Longrightarrow \sum_{i=1}^{n} \langle v_i, v \rangle \langle w_i, w \rangle = 0 \text{ for all } v \in V \text{ and } w \in W$$

This result follows by induction, with the help of 3.7.2. If all $w_i = 0$, we are done. Otherwise, by 3.7.2 the v_i are not linearly independent, say $v_1 = \sum_{i=2}^{n} \lambda_i v_i$. Then $\sum_{i=2}^{n} v_i \otimes (w_i + \lambda_i w_n) = 0$ as in 3.7.2. Applying the induction hypothesis and reversing the algebra completes the induction step.

Notation

For Hilbert spaces $\mathcal{H}$ and $\mathcal{K}$, it is conventional to use the notation $\mathcal{H} \otimes \mathcal{K}$ to denote the Hilbert space completion of the inner product space obtained as in i above from the algebraic tensor product of $\mathcal{H}$ and $\mathcal{K}$. This Hilbert space is called the **tensor product** of the Hilbert spaces $\mathcal{H}$ and $\mathcal{K}$.

3.7.4

For C^*-algebras A and B, we now consider the question of norming the $*$-algebra $A \otimes B$ so that the C^*-identity is satisfied. We will see that such a norm always exists but that, somewhat surprisingly, it need not be unique. (The lack of uniqueness does not contradict 2.1.4 because $A \otimes B$ generally will be incomplete.) One such norm, the **spatial norm**, is constructed by representing A and B as algebras of operators on Hilbert spaces $\mathcal{H}$ and $\mathcal{K}$ and using this to represent $A \otimes B$ as an algebra of operators on $\mathcal{H} \otimes \mathcal{K}$. We can describe this in more detail:

i. Choose representations $\alpha : A \to \mathcal{B}(\mathcal{H})$, $\beta : B \to \mathcal{B}(\mathcal{K})$.
ii. Show that there is a unique injection

$$\gamma : A \otimes B \to \mathcal{B}(\mathcal{H}) \otimes \mathcal{B}(\mathcal{K})$$

such that $\gamma(a \otimes b) = \alpha(a) \otimes \beta(b)$ for all $a \in A, b \in B$.

iii. Show that there is a unique injection

$$\mathcal{B}(\mathcal{H}) \otimes \mathcal{B}(\mathcal{K}) \xrightarrow{\wedge} \mathcal{B}(\mathcal{H} \otimes \mathcal{K})$$

such that $(S \otimes T)^\wedge(\xi \otimes \eta) = S\xi \otimes T\eta$ for all $S \in \mathcal{B}(\mathcal{H}), T \in \mathcal{B}(\mathcal{K})$, $\xi \in \mathcal{H}$ and $y \in \mathcal{K}$.

The desired representation of $A \otimes B$ as operators, and thus the desired norm, is obtained by composing the maps in conditions ii and iii. It can be shown the result is independent of the choices in condition i. We will have more to say about conditions i and iii later; for now we note that the existence of γ follows from the universal property of the tensor product, using the bilinear composition

$$A \times B \xrightarrow{\alpha\times\beta} \mathcal{B}(\mathcal{H}) \times \mathcal{B}(\mathcal{K}) \to \mathcal{B}(\mathcal{H}) \otimes \mathcal{B}(\mathcal{K})$$

and that the injectivity follows from 3.7.2.

Notation

In condition iii it is customary to drop the "hat" and let the notation $S \otimes T$ stand for both the element of the algebraic tensor product $\mathcal{B}(\mathcal{H}) \otimes \mathcal{B}(\mathcal{K})$ and the corresponding operator on $\mathcal{H} \otimes \mathcal{K}$.

3.7.5

As examples of naturally occurring tensor products, we mention that there are isomorphisms

$$M_n(\mathbf{C}) \otimes M_p(\mathbf{C}) \to M_{np}(\mathbf{C})$$

and, more generally, isomorphisms

$$M_n(\mathbf{C}) \otimes A \to M_n(A)$$

for any $n, p \geq 1$ and any C^*-algebra A. These are the morphisms associated with the bilinear maps

$$M_n(\mathbf{C}) \times A \to M_n(A)$$

given by $((\lambda_{ij}), a) \to (\lambda_{ij}a)$. Likewise there are injections

$$C_0(X) \otimes C_0(Y) \to C_0(X \times Y)$$

and, more generally,

$$C_0(X) \otimes A \to C_0(X, A)$$

for any locally compact Hausdorff spaces X and Y and any C^*-algebra A. It can be shown that these injections are continuous in the spatial norm of 3.7.4 and that their continuous extension to the completion of the algebraic tensor product $C_0(X) \otimes A$ is an isomorphism of the completion and $C_0(X, A)$.

Good general references for this topic are [KaR86], [Mur90] and [Tak79]; see also [Weg93]. We will return to it in Chapter 7.

3.8 CROSSED PRODUCTS

The study of dynamical systems, such as those modeling the time evolution of physical systems, has proved to be rewarding and significant. In this section we describe the role played by C^*-algebras in dynamical systems.

3.8.1

An **automorphism** of a C^*-algebra is a $*$-isomorphism $\alpha : A \to A$. For example we have the **inner automorphisms** $\alpha = \operatorname{Ad} u$, defined for unitary elements $u \in A$ by the equation $\alpha(a) = uau^*$. The set Aut A of automorphisms of A is a group under composition.

A ***C**-dynamical system** is a triple (G, α, A) consisting of a locally compact group G, a C^*-algebra A and a homomorphism α of G into Aut A, such that $g \to \alpha_g(a)$ is continuous for all $a \in A$. Such systems arise very naturally in the study of dynamical systems, where $A = C_0(X)$ and the action of G on $C_0(X)$,

$$(\alpha_g(f))(x) = f(g^{-1}x)$$

is induced by an action

$$(g, x) \to g^{-1}x : G \times X \to X$$

of G as homeomorphisms of X. The group $G = \mathbf{R}$ is the relevant one for the time evolution of physical systems. We want to consider a construction that associates a C^*-algebra to any C^*-dynamical system. One way to think of this is as an extension of the notion of the group C^*-algebra $C^*(G)$ to allow coefficients in an arbitrary C^*-algebra A.

3.8.2

In extending the group C^*-algebra construction to this situation, the idea is to enlarge the algebra A by the addition of unitary elements u_g that implement the action of G. Suppose for simplicity that G is a finite group. Then the enlarged algebra AG can be thought of as consisting of formal sums $\sum_{g\in G} a_g u_g$ with $a_g \in A$, the product and involution being determined by the requirements

$$u_g a = u_g a u_g^* u_g = \alpha_g(a) u_g$$

and

$$(au_g)^* = u_g^* a^* = u_g^* a^* u_g u_g^* = \alpha_{g^{-1}}(a^*) u_{g^{-1}}$$

Regarding the formal sums as functions $x : G \to A$, we are led to the formulas

$$(xy)(g) = \sum_h x(h)\alpha_h(y(h^{-1}g))$$

$$x^*(g) = \alpha_g(x(g^{-1})^*)$$

for all $x, y \in AG$ and $g \in G$. Notice that when α is trivial (i.e., $\alpha_g = \mathrm{id}_A$ for all g), these formulas reduce to those in 3.6.1. To norm this algebra, we again want to regard it as operators on a suitable Hilbert space. If A is represented as operators on $\mathcal{H}$, say by $\pi : A \to \mathcal{B}(\mathcal{H})$, we form the Hilbert space $\ell^2(G, \mathcal{H})$, consisting of all functions $\xi : G \to \mathcal{H}$ with inner product

$$\langle \xi, \eta \rangle = \sum_g \langle \xi(g), \eta(g) \rangle \qquad \text{for all } \xi, \eta \in \ell^2(G, \mathcal{H})$$

and define $\rho : AG \to \mathcal{B}(\ell^2(G, \mathcal{H}))$ by

$$(\rho(x)\xi)(g) = \sum_h \pi(\alpha_g^{-1}(x(h)))\xi(h^{-1}g)$$

for all $x \in AG$, $\xi \in \ell^2(G, \mathcal{H})$, and $g \in G$. Then ρ is an injective $*$-homomorphism, and the resulting C^*-algebra, written $A \rtimes_\alpha G$, is called the

crossed product or **transformation group C^*-algebra** associated to the C^*-dynamical system (A, α, G). Once again, just as in the group C^*-algebra setting, when G is infinite we have both the (full) crossed product $A \rtimes_\alpha G$ and a quotient $A \rtimes_{\alpha r} G$, the **reduced** crossed product.

3.8.3

Consider the characteristic functions δ_g of the singleton subsets $\{g\}$ of G. When A is unital, we can regard these functions as belonging to AG, and in fact $g \to \delta_g$ is then a representation of G in the unitary group of the unital C^*-algebra AG. For $a \in A$ the functions $a\delta_e$ (where e is the identity element of G) always belong to AG and $a \to a\delta_e$ is an injective $*$-homomorphism. Thus, when A is unital and G is finite, AG contains copies of both A and G. Moreover a short computation gives

$$\delta_g(a\delta_e)\delta_g^{-1} = \alpha_g(a)\delta_e \qquad \text{for all } a \in A, \ g \in G$$

so that the action of G is implemented by the unitaries δ_g. With ρ as above, $\lambda_g = \rho(\delta_g)$ defines a unitary representation of G on $\ell^2(G, \mathcal{H})$ such that

$$(\lambda_g \xi)(h) = \xi(g^{-1}h) \qquad \text{for all } \xi \in \ell^2(G, \mathcal{H}) \text{ and } g, h \in G$$

This makes sense even when A is nonunital and defines a unitary representation of G called the **left regular representation** (of multiplicity dim $\mathcal{H}$).

3.8.4

The crossed product $A \rtimes_\alpha G$ can in fact be thought of as the universal way of implementing the action of α by unitary operators. To make this precise, we need the idea of a **covariant representation** (π, u) of (A, α, G) on a Hilbert space $\mathcal{H}$, by which is meant a $*$-homomorphism $\pi : A \to \mathcal{B}(\mathcal{H})$ and a unitary representation $u : G \to \mathcal{U}(\mathcal{H})$ such that π carries α to Ad u:

$$\pi(\alpha_g(a)) = u_g \pi(a) u_g^* \qquad \text{for all } a \in A, \ g \in G$$

The universal property is that any covariant representation extends to a $*$-homomorphism $A \rtimes_\alpha G \to \mathcal{B}(\mathcal{H})$. In the general case this needs interpretation, since the crossed product contains canonical copies of A and G only when A is unital and G is discrete.

3.8.5

By way of examples, we note first that the map

$$A \times \mathbf{C}G \to AG$$

given by $(a, f) \to af$ induces an isomorphism of $A \otimes C^*(G)$ and $A \rtimes_\alpha G$ as vector spaces; when the action is trivial, this is in fact an isomorphism of $*$-algebras. Indeed these algebras are isomorphic even if α is inner, in the sense that there is a representation $g \to u_g$ of G in the unitary group of A such that $\alpha_g = \operatorname{Ad} u_g$ for all g. See [Con94, IIC] for an even more general result.

For a less trivial example, let α be an automorphism of finite order of a C^*-algebra A, say $\alpha^2 = \mathrm{id}_A$. This amounts to giving an action of $\mathbf{Z}_2 = \{\pm 1\}$ on A. Then $A \rtimes_\alpha \mathbf{Z}_2$ is closely related to the **fixed-point subalgebra** $A^\alpha = \{a \in A : \alpha(a) = a\}$. In fact, if $u \in A \rtimes_\alpha \mathbf{Z}_2$ implements α (i.e., $u = \delta_{-1}$), then by a short computation

$$p = \tfrac{1}{2}(1 + u) = \tfrac{1}{2}(\delta_1 + \delta_{-1})$$

is a projection such that

$$p(A \rtimes_\alpha \mathbf{Z}_2)p = A^\alpha p$$

The same is true of $p^\perp = \frac{1}{2}(1 - u)$. Suppose that there is an element u in the crossed product such that $u^*u = p$ and $uu^* = p^\perp$. Then the map

$$x \longrightarrow \begin{pmatrix} pxp & pxu \\ u^*xp & u^*xu \end{pmatrix}$$

is easily seen to be a $*$-isomorphism of the crossed product and $M_2(A^\alpha)$. At the other extreme, if these projections are in the center of the crossed product, then we get the direct sum of two copies of A^α. In general, the crossed product can be viewed as a subalgebra of $M_2(A)$ as follows: Let

$$A_1 = \{a \in A : \alpha(a) = a\} \quad \text{and} \quad A_{-1} = \{a \in A : \alpha(a) = -a\}$$

be the eigenspaces of α and define linear maps

$$e_1 : A \to A_1 \quad \text{and} \quad e_{-1} : A \to A_{-1}$$

by $e_{\pm 1}(a) = \frac{1}{2}(a \pm \alpha(a))$. Then

$$a \longrightarrow \begin{pmatrix} e_1(a) & e_{-1}(a) \\ e_{-1}(a) & e_1(a) \end{pmatrix}$$

embeds A in $M_2(A)$ in such a way that the action of α is given by conjugating by the unitary $\left(\begin{smallmatrix} 1 & 0 \\ 0 & -1 \end{smallmatrix}\right)$. This allows one to identify the crossed product as the subalgebra

$$\left\{ \begin{pmatrix} a & b \\ c & d \end{pmatrix} : a, d \in A_1 \text{ and } b, c \in A_{-1} \right\}$$

Similar statements are true for any finite abelian group G acting on a unital C^*-algebra A. If G is of order n, then for γ in the dual group $\hat{G}$ the elements $p_\gamma = \frac{1}{n} \sum_{g \in G} \gamma(g)\delta_g$ are pairwise orthogonal projections adding to the identity, each of which compresses the crossed product to a copy of the fixed-point algebra A^G. If, for example, $A = C^*(\hat{G})$ and the action of G is given by $(gf)(\gamma) = \gamma(g)f(\gamma)$—this is called the **dual action**—then the fixed-point algebra is $\mathbf{C}$ and the crossed product is $M_n(\mathbf{C})$, where n is the order of G. This is the algebraic case of the Takesaki and Takai duality theorems, to be discussed in Chapter 8. On the other hand, if $A = \mathbf{C}$ with the trivial action, then the projections p_γ are central and we get an n-fold direct sum:

$$\mathbf{C} \rtimes G \cong C^*(G) \cong C_0(\hat{G}) \cong \mathbf{C} \oplus \cdots \oplus \mathbf{C}$$

The best general reference for crossed product algebras is [Ped79]; see also [Tom87] and [Dav95].

3.9 *ROTATION C^*-ALGEBRAS*

The **irrational rotation algebra** $A_\theta = C(S^1) \rtimes_\theta \mathbf{Z}$ is the crossed product associated to the dynamical system $(C(S^1), \alpha, \mathbf{Z})$, where the generator of $\mathbf{Z}$ acts on $C(S^1)$ by rotating the circle S^1 through the angle $2\pi\theta$ and θ is an irrational number in $(0, 1)$. If we let $C(S^1)$ act on $L^2(S^1)$ as multiplication operators, then A_θ can be simply described as the C^*-subalgebra of $\mathcal{B}(L^2(S^1))$ generated by $C(S^1)$ and the unitary operator U defined by $(U\xi)(z) = \xi(\lambda z)$ for all $\xi \in L^2(S^1)$ and $z \in S^1$, where $\lambda = \exp(2\pi i\theta)$. If V is the unitary given by multiplication by z, then $UV = \lambda VU$ and A_θ can be described as the **universal C^*-algebra** with unitary generators satisfying this relation. This means that if A is any C^*-algebra generated by unitary elements u and v such that $uv = \lambda uv$, then there is a morphism $A_\theta \to A$ taking U to u and V to v.

These C^*-algebras have attracted a good deal of attention, and a number of interesting things are known about them. For example, the A_θ are simple (so that the morphism just mentioned is in fact an isomorphism), and except for $A_\theta \cong A_{1-\theta}$ (obtained by sending U to V and V to U), they are pairwise nonisomorphic.

The proof of simplicity is rather technical, but the main ideas can be outlined rather easily. Since A_θ is generated by U and $C(S^1)$, it can be regarded as the completion of the algebra of "Laurent polynomials" $\sum_{k=-n}^{k=n} f_k u^k$, computation with these being carried out according to the rules $u^k f = \alpha_k(f)u^k$ and $(fu^k)^* = \alpha_{-k}(f)u^{-k}$. The first fact that one needs is that the map

$$\sum_{k=-n}^{k=n} f_k u^k \longrightarrow f_0$$

extends (by continuity) to define a map $E : A_\theta \to C(S^1)$ that is a **conditional expectation**. Briefly this means that E is a unital positive linear contraction with the module property $E(fx) = fE(x)$ for all $f \in C(S^1)$ and $x \in A_\theta$ (see 9.5.3 below). One also needs the following additional properties of E:

i. $E(uxu^*) = \alpha(E(x))$ for all $x \in A_\theta$.
ii. $E(x^*x) = 0$ only for $x = 0$.
iii. $E(J) \subset J$ for any ideal J of A_θ.

Suppose θ is rational. Then there is a proper closed set $F \subset S^1$ which is invariant; that is, $\lambda F = F$, and the closure of the "polynomials" (see above) having all $f_k = 0$ on F is then a proper ideal J of A_θ. It is proper because

$$\left\|1 - \sum f_k u^k\right\| \geq \left\|E(1 - \sum f_k u^k)\right\| = \|1 - f_0\| \geq 1$$

and so $\|1 - x\| \geq 1$ for all $x \in J$. On the other hand, if J is any ideal of A_θ, then by property iii and the module property $E(J)$ is an ideal of $C(S^1)$, so there is a closed set $F \subset S^1$ such that $E(J)$ consists of all functions vanishing on F. Moreover $\alpha(E(J)) = E(J)$ by property i, and it follows that F is invariant. Assuming now that θ is irrational, we conclude that F is empty or all of S^1, hence that $E(J)$ is all of $C(S^1)$ or $\{0\}$. In the first case $1 \in J$ by property iii, so $J = A_\theta$, and in the second case $J = 0$ by property ii.

References: [Dav95]; [EfH67]; [Pow78]; [Put90]; [Rie81]; [Tom87].

3.10 DIRECT LIMITS

A **direct** (or **inductive**) **system** of C^*-algebras, indexed by an upward-directed set I, is a family of C^*-algebras A_i and $*$-homomorphisms $\varphi_{ij} : A_i \to A_j$ for $i < j$ such that $\varphi_{jk}\varphi_{ij} = \varphi_{ik}$ for $i < j < k$. The **direct** (or **inductive**) **limit** is the C^*-algebra $A = \lim_{\to} A_i$, defined as follows: Let A_0 be the $*$-subalgebra of the product $\prod A_i$ consisting of all $a = (a_i)$ for which there exists i_0 such that $\varphi_{ij}(a_i) = a_j$ for all $i_0 < i < j$. Then $\|a\| = \lim \|a_i\|$ exists by 2.1.5 and is a seminorm that satisfies the C^*-identity. The direct limit is then the completion of A_0, modulo the ideal of elements of norm zero.

Note that there are $*$-homomorphism $\varphi_i : A_i \to A$ that are compatible with the φ_{ij}, in the sense that $\varphi_j\varphi_{ij} = \varphi_i$ for all $i < j$. In fact φ_i is induced by the map taking $a \in A_i$ to $(a_j) \in A_0$ where $a_i = a, a_j = \varphi_{ij}(a)$ for $i < j$, and $a_j = 0$ otherwise.

The direct limit algebra A is determined up to isomorphism by a universal property: For any C^*-algebra B and morphisms $\psi_i : A_i \to B$ compatible with the φ_{ij}, there is a unique morphism $A \to B$ such that composition with the above maps $\varphi_i : A_i \to A$ gives the ψ_i.

The most elementary direct systems are the **direct sequences**

$$A_1 \to A_2 \to A_3 \to \cdots$$

especially when the maps of the system are inclusions (of subalgebras of a fixed C^*-algebra). In this case the direct limit can be identified as the norm closure of $\bigcup A_n$. In general $\bigcup \varphi_i(A_i)$ is dense in the direct limit algebra.

3.10.1

It is not difficult to see that if all the A_i are simple, so is A. (It is equivalent to say that any morphism $A \to B$ is either zero or injective, and this results from the fact that the compositions $A_i \to A \to B$ are all either zero or eventually injective.) In particular, this is the case when the A_i are matrix algebras. A C^*-algebra is **uniformly hyperfinite** (**UHF**) if it is the direct limit of a sequence

$$M_{n_1}(\mathbf{C}) \to M_{n_2}(\mathbf{C}) \to M_{n_3}(\mathbf{C}) \to \cdots$$

of matrix algebras and unital maps. The latter condition forces each n_{i+1} to be a multiple of n_i and the maps to be inflations

$$a \longrightarrow \begin{pmatrix} a & & 0 \\ & \ddots & \\ 0 & & a \end{pmatrix}$$

up to unitary equivalence. Observe that UHF algebras can be regarded as "infinite tensor products" of matrix algebras, in the sense that they are direct limits of the (finite) partial products. This results from the fact that if p divides n, say $n = pq$, then $M_n(\mathbf{C}) \cong M_p(\mathbf{C}) \otimes M_q(\mathbf{C})$. In this picture the inflation map becomes

$$a \to a \otimes 1 : M_p(\mathbf{C}) \to M_p(\mathbf{C}) \otimes M_q(\mathbf{C})$$

These algebras, introduced by Glimm [Gli60], have received a great deal of attention. Glimm proved that a complete isomorphism invariant is given by the sequence $e_1, e_2, \ldots$ of extended nonnegative integers $0 \leq e_i \leq +\infty$, where e_i is the maximum multiplicity of the ith prime p_i in any n_j. The corresponding algebra is usually denoted $U(n)$, where n is the "supernatural number" $p_1^{e_1} p_2^{e_2} \ldots$. He also obtained an internal characterization: The UHF algebras are the separable unital C^*-algebras A for which, given any finite set of elements $a_1, \ldots, a_k \in A$ and $\varepsilon > 0$, there exists a C^*-subalgebra B of A such that B contains the unit of A, $B \cong M_n(\mathbf{C})$ for some n and B contains norm approximants of the a_i to within ε.

An interesting example is the **CAR** or **Fermion algebra** of mathematical physics, so-called because of its use in the study of representations of the canonical anticommutation relations for systems of particles obeying Fermi statistics. It is the algebra $U(2^\infty)$, specified by its invariant 2^∞ as the direct limit of

$$M_2(\mathbf{C}) \to M_2(\mathbf{C}) \otimes M_2(\mathbf{C}) \to \cdots$$

We will indicate later how the Powers factors [Pow67], a continuum of nonisomorphic type III factors, can be constructed from the CAR algebra.

3.10.2

In a conceptually significant step forward, Dixmier [Dix67] exploited ideas of Murray and von Neumann, which they had developed for their classification of von Neumann algebras ([MvN36]; see Chapter 4 below), to extend Glimm's results on UHF algebras to **matroid algebras**. These are defined in the same way as UHF algebras, except that the maps in the direct sequence are not required to be unital. An elementary example, not UHF, is the algebra $\mathcal{K}$ of compact operators, which can be obtained as the limit of

$$M_1(\mathbf{C}) \to M_2(\mathbf{C}) \to M_3(\mathbf{C}) \to \cdots$$

with maps $a \to \begin{pmatrix} a & 0 \\ 0 & 0 \end{pmatrix}$. This is easily seen by identifying $M_n(\mathbf{C})$ with $\mathcal{B}(\mathbf{C}^n)$ and $\mathbf{C}^n$ with the subspace of ℓ^2 spanned by the first n basis vectors.

The Murray-von Neumann procedure, as applied by Dixmier to C^*-algebras, attaches to any C^*-algebra A an abelian partial semigroup $D(A)$, as follows. Let $P(A)$ be the set of projections in A, and introduce in $P(A)$ the equivalence relation $p \sim q$ if there exists $u \in A$ with $u^*u = p$ and $uu^* = q$. Then $D(A)$ is the set $P(A)/\sim$ of equivalence classes, with the sum of two equivalence classes being defined as

$$[p] + [q] = [p + q]$$

if they have orthogonal representatives p and q; otherwise, the sum is left undefined. The idea behind this **Murray-von Neumann equivalence** is that equivalent projections ("subspaces") should have the same dimension (relative to A) and that this dimension should take its values in $D(A)$. For example, $D(M_n(\mathbf{C}))$ is isomorphic to $\{0, 1, \ldots, n\}$ with ordinary addition (the sum of k and m is $k + m$ if $k + m \leq n$ and otherwise is not defined).

Returning to the case of matroid algebras, Dixmier obtained a complete isomorphism invariant (n, θ), where n is a supernatural number as in the preceding section and θ is an extended nonnegative real number. The connection between a matroid algebra and its invariant is that

$$D(A) \cong Q(n) \cap [0, \theta]$$

where $Q(n)$ is the additive subgroup of $\mathbf{Q}$ consisting of all p/q with q a divisor of n. Thus, for example, $\mathcal{K}$ has invariant $(1, +\infty)$ and $D(\mathcal{K}) \cong \mathbf{Z}_+$. Using this classification, Dixmier showed that matroid algebras fall into three classes: UHF (if unital); subalgebras of UHF algebras U of the form $(\bigcup e_n U e_n)^-$, where $\{e_n\}$ is an increasing sequence of projections in U; and algebras of the form UHF $\otimes \mathcal{K}$. He also obtained an internal characterization of matroid algebras analogous to Glimm's result for UHF algebras.

This classification was subsequently extended by Elliott [Ell76] to limits of direct sequences of arbitrary finite-dimensional C^*-algebras (the AF algebras; see below), but the power of the method has only recently been fully appreciated. The "Elliott program" proposes to use these and related invariants to classify large families of C^*-algebras, perhaps eventually even all separable amenable ones. Already striking progress toward this goal has been recorded [Ell93; Ell95]. We will return to this in Chapter 10.

3.10.3

Bunce and Deddens [BuD75] studied the interesting class of C^*-algebras, analogous to the UHF algebras and now called **Bunce-Deddens algebras**, obtained by replacing $M_n(\mathbf{C})$ by $M_n(C(S^1))$. They are simple and have the same classification theorem as the UHF algebras.

There is a connection with weighted shift operators. Let ℓ^2 be the usual space of one-way sequences, and recall that an operator T on ℓ^2 is a **weighted shift** if $T = SD$, where S is the forward shift and D is diagonal, both relative to the standard orthonormal basis of ℓ^2. The **weights** of T are the diagonal entries of D. For any $n \geq 1$, let $A(n)$ be the C^*-subalgebra of $\mathcal{B}(\ell_2)$ generated by all weighted shifts with weight sequence having period n. Then $A(n)$ is generated by S, and the diagonal unitary $V = \text{diag}(1, \lambda, \lambda^2, \ldots)$ with λ a primitive nth root of unity. Moreover $M_n(C(S^1))$ is generated by

$$u_n = \begin{pmatrix} 0 & z \\ I_{n-1} & 0 \end{pmatrix} \quad \text{and} \quad v_n = \text{diag}(1, \lambda, \ldots, \lambda^{n-1})$$

and there is a short exact sequence

$$0 \to \mathcal{K}(\ell^2) \hookrightarrow A(n) \xrightarrow{\pi} M_n(C(S^1)) \to 0$$

with $\pi(S) = u_n$ and $\pi(V) = v_n$. When $n = 1$, this is the Toeplitz extension, which is described in detail in Section 3.11 below.

Suppose now that n divides m, say $m = nk$, so that $A(n)$ is a subalgebra of $A(m)$. What is the morphism $M_n(C(S^1)) \to M_m(C(S^1))$ induced by this inclusion? Of course it takes u_n to u_m, and it is not difficult to see that it is the k-fold inflation $a \to a \oplus \ldots \oplus a$ on the constant matrices. In particular, it takes $zI_n = u_n^n$ to u_m^n and hence $z^k I_n$ to $u_m^m = zI_m$. If $\{n_k\}$ is a sequence of positive integers with each dividing the next, then the corresponding Bunce-Deddens algebra is the direct limit of

$$M_{n_1}(C(S^1)) \to M_{n_2}(C(S^1)) \to \cdots$$

with morphisms as described above. Alternatively, the Bunce-Deddens algebra can be described as the quotient $(\bigcup A(n_k))^{-}/\mathcal{K}$.

With the work of Glimm, Dixmier, Elliott, Bunce and Deddens in mind, Effros [Eff82] put forward the problem of finding suitable invariants for studying limits of direct sequences of the form

$$C(X_1) \otimes A_1 \to C(X_2) \otimes A_2 \to \cdots$$

where the X_n are CW-complexes and the A_n are finite-dimensional C^*-algebras. Such direct limit algebras are called **approximately homogeneous (AH)** [Bla93].

3.10.4

As mentioned above, by an approximately finite-dimensional or **AF algebra** is meant the direct limit of a direct sequence of finite-dimensional C^*-algebras. These algebras were introduced and studied by Bratteli [Bra72] who, in particular, introduced the Bratteli diagrams (1.1.3) to encode the direct system. For example, the diagram (drawn horizontally)

$$2 \;\text{——}\; 4 \;\text{——}\; 8 \;\text{——}\; \cdots$$

represents $M_2(\mathbf{C}) \to M_4(\mathbf{C}) \to M_8(\mathbf{C}) \to \cdots$ with multiplicity 1 embeddings; the direct limit is $\mathcal{K}$. The diagram

$$2 \;=\!=\!=\; 4 \;=\!=\!=\; 8 \;=\!=\!=\; \cdots$$

represents the same sequence of algebras but with multiplicity 2 embeddings. Its direct limit is the UHF algebra $U(2^\infty)$, as is the limit of the direct sequence represented by

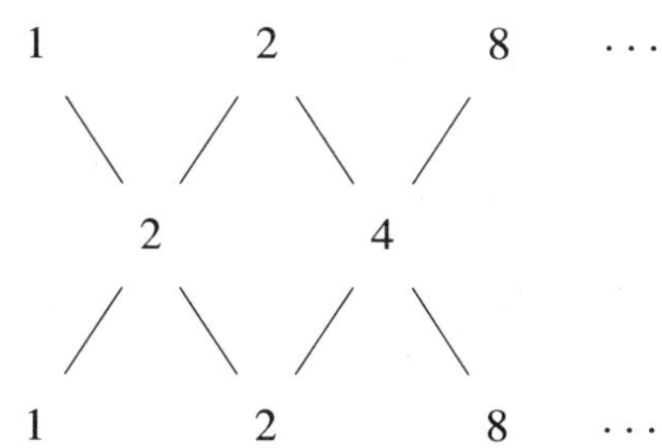

Finally, the infinite binary tree (to be read vertically)

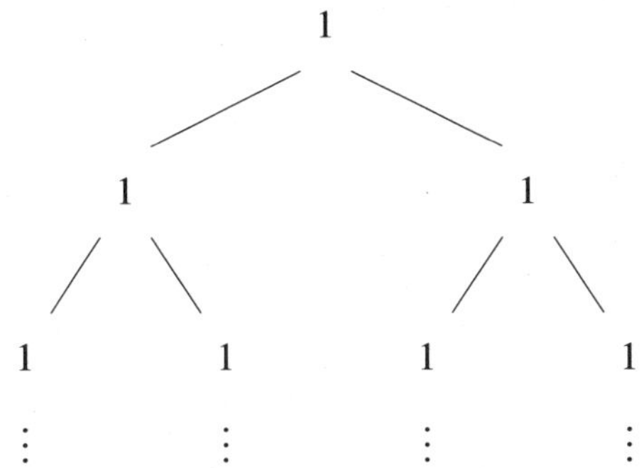

shows that the algebra $C(K)$ of continuous functions on the Cantor set K is an AF algebra.

Among other results, Bratteli proved that if direct limits $A = \lim_{\rightarrow} A_n$ and $B = \lim_{\rightarrow} B_n$ have the same Bratteli diagram, then $A \cong B$. The hypothesis implies that one has a diagram

$$\begin{array}{ccccccc} A_1 & \longrightarrow & A_2 & \longrightarrow & A_3 & \longrightarrow & \cdots \\ \downarrow & & \downarrow & & \downarrow & & \\ B_1 & \longrightarrow & B_2 & \longrightarrow & B_3 & \longrightarrow & \cdots \end{array}$$

in which the vertical maps are isomorphisms. It need not be commutative, but the hypothesis means that the vertical maps can be adjusted, one at a time (beginning with the second one) by inner automorphisms of the B_i, to make it commutative. Then the compositions $A_n \rightarrow B_n \rightarrow B$ form a compatible family that induces (by the universal property of the direct limit) the desired morphism $A \rightarrow B$. That it is an isomorphism follows by reversing the roles of the direct sequences.

Bratteli also obtained an analogue of the internal characterization of the UHF and matroid cases.

3.11 THE TOEPLITZ ALGEBRA

Let S be the forward unilateral shift operator, defined on the space $\ell^2 = \ell^2(\mathbf{N})$ of one-way sequences as in 3.10.3. The **Toeplitz algebra** is the C^*-subalgebra

$C^*(S)$ of $\mathcal{B}(\ell^2)$ generated by S. We will see that this algebra plays an important role in the K-theory of C^*-algebras.

3.11.1 *Theorem* [Cob67]

There is a short exact sequence

$$0 \to \mathcal{K}(\ell^2) \hookrightarrow C^*(S) \xrightarrow{\pi} C(\partial\mathbf{D}) \to 0$$

with $\pi(S) = z$.

This sequence is called the **Toeplitz extension.** We are denoting by $\mathcal{K}(\ell^2)$ the compact operators on ℓ^2 and by $\partial\mathbf{D}$ the circle S^1, identified as the boundary of the unit disc $\mathbf{D}$ in the complex plane. We will sketch two proofs. The idea of the first is to show that $C^*(S)$ acts irreducibly on ℓ^2 (i.e., has no proper closed invariant subspaces) and then use the following general fact:

3.11.2 *Lemma*

If $\mathcal{A}$ *is an irreducible* C^**-subalgebra of* $\mathcal{B}(\mathcal{H})$ *such that* $\mathcal{A} \cap \mathcal{K}(\mathcal{H}) \neq \{0\}$, *then* $\mathcal{A} \supset \mathcal{K}(\mathcal{H})$.

Proof

The hypothesis implies, by elementary spectral theory, that $\mathcal{A}$ contains a nonzero projection of finite rank, and hence a minimal one, say E. By minimality $E\mathcal{A}E = \mathbf{C}E$, again using spectral theory. It follows that rank $E = 1$. To see this, fix a unit vector $\xi_0 \in E\mathcal{H}$. Then for any $\eta \in E\mathcal{H}$ orthogonal to ξ_0, we have

$$\langle A\xi_0, \eta\rangle = \langle AE\xi_0, E\eta\rangle = \langle EAE\xi_0, \eta\rangle = \langle \lambda E\xi_0, \eta\rangle = 0$$

for all $A \in \mathcal{A}$, since $EAE \in \mathbf{C}E$. But $\overline{\mathcal{A}\xi_0}$ is a nonzero invariant subspace for $\mathcal{A}$, so by hypothesis $\overline{\mathcal{A}\xi_0} = \mathcal{H}$, $\eta = 0$, and rank $E = 1$ as asserted.

The proof of the lemma is completed by showing that $\mathcal{A}$ contains all projections of rank one. These have the form

$$F\xi = \langle \xi, \zeta\rangle\zeta \qquad \text{for all } \xi \in \mathcal{H}$$

for some unit vector ζ. Since $\overline{\mathcal{A}\xi_0} = \mathcal{H}$, there is a sequence $A_n \in \mathcal{A}$ with $A_n\xi_0 \to \zeta$. Then

$$\|(A_n E A_n^* - F)\xi\| = \|\langle A_n^*\xi, \xi_0\rangle A_n\xi_0 - \langle \xi, \zeta\rangle\zeta\|$$

$$\leq \|\langle \xi, A_n\xi_0 - \zeta\rangle A_n\xi_0\| + \|\langle \xi, \zeta\rangle (A_n\xi_0 - \zeta)\|$$

$$\leq \|\xi\| \|A_n\xi_0 - \zeta\| (\|A_n\xi_0\| + 1)$$

and therefore $F \in \mathcal{A}$.

Proof of the Theorem

To prove that $C^*(S)$ contains $\mathcal{K}$, note first that P_0, the projection on the space spanned by $e_0 = (1, 0, 0, \cdots)$, is in $C^*(S)$ since $P_0 = I - SS^*$. Now for any $\xi = (\xi_0, \xi_1, \cdots) \in \ell^2$ and $n \in \mathbf{N}$ we have $P_0 S^{*n}\xi = \xi_n e_0$. It follows that any nonzero invariant subspace contains e_0, hence all $e_n = S^n e_0$, hence all of ℓ^2. By the lemma $C^*(S)$ contains the compact operators, obviously as an ideal.

It remains to consider the quotient map π. The polynomials in S and S^* are dense in $C^*(S)$ and $S^*S - SS^* = P_0$ is compact, so $A = C^*(S)/\mathcal{K}$ is commutative. It is a C^*-algebra by 2.5.4, and consequently the Gelfand transform $A \to C(\hat{A})$ is an isometric $*$-isomorphism by 2.2.2. The coset $s = S + \mathcal{K}$ generates A, and this implies that $\varphi \to \varphi(s)$ is a homeomorphism of $\hat{A}$ with $\sigma(s)$. It is easily checked that if $\hat{A}$ is replaced by $\sigma(s)$, then the Gelfand transform $A \to C(\sigma(s))$ so modified takes the generator s to the function z. To complete the proof, we need $\sigma(s) = \partial\mathbf{D}$. This follows from **Atkinson's theorem** [Mur90, p. 28] which asserts that for any $T \in \mathcal{B}(\mathcal{H})$, its coset $T + \mathcal{K}$ in the **Calkin algebra** $\mathcal{C} = \mathcal{B}(\mathcal{H})/\mathcal{K}(\mathcal{H})$ is invertible if and only if T is Fredholm. Thus

$$\sigma_{\mathcal{C}}(s) = \{\lambda : S - \lambda \text{ is not Fredholm}\}$$

and the latter is readily seen to be $\partial\mathbf{D}$. But $\sigma_{\mathcal{C}}(s) = \sigma(s)$ by 2.1.8, since A is a unital C^*-subalgebra of $\mathcal{C}$.

3.11.3

There is another proof of 3.11.1 [Cob69; Dou72] that sheds considerable light on the internal structure of the Toeplitz algebra. Recall that the Hardy space H^2 is the subspace of $L^2(\partial\mathbf{D})$ having the exponentials $e^{in\theta}$ for $n \geq 0$ as an orthonormal basis. The shift operator is therefore realized on H^2 as multiplication by $e^{i\theta}$. More generally a **Toeplitz operator** is any operator on H^2 of the form $T_\varphi = PM_\varphi$, where $\varphi \in L^\infty(\partial\mathbf{D})$, M_φ is multiplication by φ, and P is the projection of L^2 on H^2. If φ is a polynomial in $e^{i\theta}$ and $e^{-i\theta}$, then T_φ is the same polynomial in S and S^*, and so $T_\varphi \in C^*(S)$. Since $\|T_\varphi\| = \|\varphi\|_\infty$ for all $\varphi \in L^\infty$, it follows that $T_\varphi \in C^*(S)$ whenever $\varphi \in C(\partial\mathbf{D})$. The key technical point is the inequality

$$\|T_\varphi + K\| \geq \|T_\varphi\| \qquad \text{for all } \varphi \in C(\partial\mathbf{D}),\ K \text{ compact}$$

It follows from this that each element of $C^*(S)$ is uniquely of the form $T_\varphi + K$ for some $\varphi \in C(\partial\mathbf{D})$ and $K \in \mathcal{K}(H^2)$. The **symbol map** $T_\varphi + K \to \varphi$ is then well-defined and is the map $\pi : C^*(S) \to C(\partial\mathbf{D})$ of the theorem.

3.11.4

The shift operator S is a proper isometry; equivalently, it satisfies $S^*S = I$ and is nonunitary. A further characterization of the Toeplitz algebra is that it is the universal C^*-algebra with a single generator S subject to the relation $S^*S = 1$ [Cob69; Mur90]. Algebras generated by two or more isometries have been studied by Cuntz [Cun77; Cun81]. For $n \geq 2$ the **Cuntz algebra** $\mathcal{O}_n$ is the universal C^*-algebra generated by isometries $u_1, u_2, \cdots, u_n$ such that $u_1u_1^* + \cdots u_nu_n^* = 1$. These algebras are known to be simple and pairwise non-isomorphic; they play a fundamental role in the study of simple C^*-algebras [Rør94].

3.11.5

In addition to the Toeplitz extension, one might ask what other extensions (see 1.3) of $C(\partial\mathbf{D})$ by $\mathcal{K}$ there are. The answer [BDF73] is that index considerations allow one to define a bijection

$$\mathrm{Ext}(C(\partial\mathbf{D}), \mathcal{K}) \to \mathbf{Z}$$

which is in fact an isomorphism of groups for a suitable operation on extensions. Namely the sum of extensions

$$0 \to \mathcal{K}(\mathcal{H}_i) \to \mathcal{A}_i \xrightarrow{\pi_i} C(\partial\mathbf{D}) \to 0$$

for $i = 1, 2$ is the extension given by

$$\mathcal{A} = \left\{ \begin{pmatrix} S_1 & K_{12} \\ K_{21} & S_2 \end{pmatrix} : S_i \in \mathcal{A}_i,\ K_{ij} \in \mathcal{K}(\mathcal{H}_j, \mathcal{H}_i),\ \pi_1(S_1) = \pi_2(S_2) \right\}$$

$$\pi \begin{pmatrix} S_1 & K_{12} \\ K_{21} & S_2 \end{pmatrix} = \pi_1(S_1) = \pi_2(S_2)$$

The kernel of π is $\mathcal{K}(\mathcal{H}_1 \oplus \mathcal{H}_2)$; we are implicitly making use of the fact that the algebra of compact operators is independent (up to isomorphism) of the particular separable infinite-dimensional Hilbert space. The zero element is defined by

$$\mathcal{A}_0 = \{M_\varphi + K : \varphi \in C(\partial\mathbf{D}),\ K \text{ compact}\}$$

$$\pi_0(M_\varphi + K) = \varphi.$$

Note that $\mathcal{A}_0$ is a C^*-subalgebra of $\mathcal{B}(L^2(\partial\mathbf{D}))$ by 2.5.6. The integer associated to an extension

$$0 \to \mathcal{K} \to \mathcal{A} \xrightarrow{\pi} C(\partial\mathbf{D}) \to 0$$

is simply the Fredholm index of any $T \in \mathcal{A}$ with $\pi(T) = z$.

These ideas have been extended from the circle to arbitrary compact spaces [BDF77; Dou80] and beyond; we will return to this topic in Chapter 10.

3.12 MULTIPLIER ALGEBRAS AND EXTENSIONS

A **double centralizer** of a C^*-algebra A is a pair (L, R) of maps from A into A such that [Joh64]

$$R(a)b = aL(b) \qquad \text{for all } a, b \in A$$

As an example, let A be an ideal in a C^*-algebra B, and take L_b and R_b to be left and right multiplication on A by any fixed element $b \in B$. In this case the double centralizer condition is just the associativity of multiplication in B: If $a, c \in A$ and $b \in B$, then

$$R_b(a)c = (ab)c = a(bc) = aL_b(c)$$

3.12.1

It follows from the double centralizer condition that the maps L and R are necessarily bounded, linear and satisfy

$$L(ab) = L(a)b \quad \text{and} \quad R(ab) = aR(b) \qquad \text{for all } a, b \in A$$

For example, for the additivity of L, we compute

$$\begin{aligned} xL(a+b) = R(x)(a+b) &= R(x)a + R(x)b \\ &= xL(a) + xL(b) = x(L(a)+L(b)) \end{aligned}$$

and use the fact that if $xc = 0$ for all x in a C^*-algebra, then in particular $c^*c = 0$ and so $c = 0$. The other assertions follow in a similar manner, with the help of the closed graph theorem in the case of boundedness.

3.12.2

The **multiplier algebra** of a C^*-algebra A is the set $M(A)$ of its double centralizers. Its linear structure is inherited as a subset of the product $\mathcal{B}(A) \times \mathcal{B}(A)$, multiplication is given by

$$(L, R)(L', R') = (LL', R'R)$$

and the involution by

$$(L, R)^* = (R^*, L^*)$$

where for any $T : A \to A$, T^* is defined by $T^*(a) = (T(a^*))^*$. For the norm let

$$\|(L, R)\| = \|L\| = \|R\|$$

and note that the latter equality results from

$$\begin{aligned} \|L(b)\| &= \sup\{\|aL(b)\| : \|a\| \leq 1\} \\ &= \sup\{\|R(a)b\| : \|a\| \leq 1\} \leq \|R\|\|b\| \end{aligned}$$

and a similar computation reversing the roles of L and R.

Theorem [Bus68]

With the operations and norm defined above, $M(A)$ is a C^*-algebra.

Proof

We check only the C^*-identity. Let $T = (L, R)$ be a double centralizer, so that $T^*T = (R^*L, RL^*)$. Then

$$\begin{aligned}\|T\|^2 = \|L\|^2 &= \sup\{\|L(a)\|^2 : \|a\| \leq 1\}\\ &= \sup\{\|L(a)^*L(a)\| : \|a\| \leq 1\}\\ &= \sup\{\|L^*(a^*)L(a)\| : \|a\| \leq 1\}\\ &= \sup\{\|R(L^*(a^*))a\| : \|a\| \leq 1\}\\ &\leq \|RL^*\| = \|T^*T\|\end{aligned}$$

3.12.3

Here are some of the more important properties of $M(A)$:

i. $a \to (L_a, R_a)$ is an isometric $*$-homomorphism $A \to M(A)$ with range a two-sided **essential ideal** (i.e. $Am = 0$ or $mA = 0$ occurs only for $m = 0$) of $M(A)$. It is customary to identify A with its image in $M(A)$.

ii. $M(A)$ has an identity, namely (I, I).

iii. If A is unital then $A = M(A)$. In other words, the map $A \to M(A)$ of property i is surjective. In fact, if $(L, R) \in M(A)$, then for all $x \in A$,

$$L(x) = L(1 \cdot x) = L(1)x$$

and so $(L, R) = (L_a, R_a)$ with $a = L(1) = R(1)$.

iv. If A is commutative, so is $M(A)$.

v. $M(A)$ is maximal among C^*-algebras containing A as an ideal in the following sense: if B is any C^*-algebra containing A as a closed ideal, then the map $A \to M(A)$ of property i extends uniquely to a $*$-homomorphism $B \to M(A)$. Moreover, if A is essential in B, then the extension is injective, and if B is unital, so is the extension.

We indicate how the proof of v goes. For existence take the map $b \to (L_b, R_b)$. For uniqueness, for any extension ψ and for any $b \in B$ and $a \in A$, if $\psi(b) = (L, R)$, then

$$(L_{ab}, R_{ab}) = \psi(ab) = \psi(a)\psi(b) = (L_a, R_a)(L, R) = (L_aL, RR_a)$$

Thus $aL(x) = L_aL(x) = L_{ab}(x) = abx$; hence $L(x) = bx$ for all $x \in A$ so that $L = L_b$. The multiplier identity now gives $R = R_b$ as well.

3.12.4 Examples

We will see later that if $\mathcal{A}$ is a C^*-subalgebra of $\mathcal{B}(\mathcal{H})$, and if $\mathcal{A}$ acts **nondegenerately**, in the sense that $\mathcal{A}\xi = \{0\}$ occurs only for $\xi = 0$, then

$$M(\mathcal{A}) \cong \{T \in \mathcal{B}(\mathcal{H}) : T\mathcal{A}, \mathcal{A}T \subset \mathcal{A}\}$$

in such a way that the canonical embedding $\mathcal{A} \to M(\mathcal{A})$ becomes inclusion. Since, as has already been remarked, we are going to see that any C^*-algebra can be realized in this fashion, the above result can be used in principle to determine any multiplier algebra. Thus

EXAMPLE 1. $M(\mathcal{K}(\mathcal{H})) = \mathcal{B}(\mathcal{H})$, where $\mathcal{K}(\mathcal{H})$ is the algebra of compact operators on $\mathcal{H}$. This results from the fact that $\mathcal{K}(\mathcal{H})$ is an ideal in $\mathcal{B}(\mathcal{H})$.

EXAMPLE 2. $M(C_0(X)) = C_b(X)$, the algebra of continuous bounded functions on the locally compact Hausdorff space X. This follows from the above by regarding $C_0(X)$ as operators on $\mathcal{H} = \ell^2(X)$, acting by multiplication. In particular $M(c_0) = \ell^\infty$.

3.12.5

Let βX be the Stone-Čech compactification of the locally compact Hausdorff space X. Recalling that $C_b(X) \cong C(\beta X)$, we see that the multiplier algebra can be viewed as a noncommutative generalization of the Stone-Čech compactification. This idea needs to be used with caution. For example, since any map $X \to K$ into a compact space extends to a map $\beta X \to K$, it follows that β is a (covariant) functor; however, M is not. Thus the inclusion $c_0 \to c$ does not extend to $M(c_0) \to M(c)$ (i.e., $\ell^\infty \to c$). To see this one can use $\ell^\infty \cong C(\beta\mathbf{N})$, $c \cong C(\mathbf{N} \cup \{\infty\})$, and the fact that morphisms $C(\beta\mathbf{N}) \to C(\mathbf{N} \cup \{\infty\})$ are induced by continuous maps $\mathbf{N} \cup \{\infty\} \to \beta\mathbf{N}$.

In some useful cases maps $A \to B$ do extend to $M(A) \to M(B)$. For example, this is the case if the image of A contains an approximate unit for B [Ped79].

3.12.6

Busby [Bus68] studied multipliers from the point of view of extensions of algebras. The connection arises from the map

$$\operatorname{Hom}(C', C) \times \operatorname{Ext}(C, A) \to \operatorname{Ext}(C', A)$$

defined by the pullback construction: Given a morphism $C' \to C$ and an extension $0 \to A \to B \to C \to 0$, there is an extension $0 \to A \to B' \to C' \to 0$, unique up to equivalence, that fits in the commutative diagram

$$\begin{array}{ccccc} A & \longrightarrow & B' & \longrightarrow & C' \\ \| & & \downarrow & & \downarrow \\ A & \longrightarrow & B & \longrightarrow & C \end{array}$$

Consider the extension

$$0 \to A \to M(A) \to C(A) \to 0$$

where $C(A)$ is the **corona algebra** $M(A)/A$ (so-called because if $A = C_0(X)$, then $C(A) = C(\beta X \setminus X)$). From the above this gives a map

$$\operatorname{Hom}(C', C(A)) \to \operatorname{Ext}(C', A)$$

and one of Busby's main results (Th. 4.3) is that it is a bijection. In particular, extensions of $\mathcal{K}(\mathcal{H})$ by a C^*-algebra C correspond to morphisms from C into the Calkin algebra $\mathcal{C}(\mathcal{H}) = \mathcal{B}(\mathcal{H})/\mathcal{K}(\mathcal{H})$.

3.12.7

An extension $0 \to A \to B \to C \to 0$ is **split** if there is a morphism $C \to B$ such that the composition $C \to B \to C$ is the identity. It is not difficult to prove that split extensions correspond in 3.12.6 to morphisms $C \to C(A)$ that factor through $M(A)$ [Bus68, 5.3]:

$$\begin{array}{ccc} & \nearrow & M(A) \\ & & \downarrow \\ C & \longrightarrow & C(A) \end{array}$$

The Toeplitz extension 3.11.1 corresponds to the map $C(\partial\mathbf{D}) \to \mathcal{C}(\ell^2(\mathbf{N}))$ defined by taking the generator z of $C(\partial\mathbf{D})$ to the image s of the unilateral shift S in the Calkin algebra $\mathcal{C}(\ell^2(\mathbf{N}))$. Now no preimage T of s in $\mathcal{B}(\ell^2(\mathbf{N}))$ can be unitary (because, for example, index T = index $S = -1$), and therefore the Toeplitz extension does not split.

3.13 GROUPOID C^*-ALGEBRAS

One can attach C^*-algebras to certain grouplike objects, called **groupoids**, in a way that extends the construction of group C^*-algebras and transformation group C^*-algebras. This has proved to be significant, for example, in the study of noncommutative differential geometry [Con94], where the groupoid plays an intermediate role allowing one to encode geometric information in a C^*-algebra. The basic reference for groupoids is [Ren80].

3.13.1

A groupoid may be described concisely as a small category in which every morphism is an isomorphism. Thus it consists of a set A (of "arrows"), a subset U (of "units"), maps $d, r : A \to U$ (the "domain" and "range"), an associative partially defined product $(a, b) \to ab$ on A ("composition" of arrows) and an involution $a \to a^{-1}$ on A ("reverse arrow") such that

i. $d(u) = r(u) = u$ for all $u \in U$;
ii. if $a, b \in A$ and $d(a) = r(b)$, then ab is defined, $d(ab) = d(b)$ and $r(ab) = r(a)$;
iii. $ad(a) = a = r(a)a$ for all $a \in A$;
iv. $d(a^{-1}) = r(a)$ and $a^{-1}a = d(a)$ for all $a \in A$.

Of course it follows from property iv, by replacing a by a^{-1}, that $r(a^{-1}) = d(a)$ and $aa^{-1} = r(a)$ for all $a \in A$.

A groupoid is **principal** if the map $(d, r) : A \to U \times U$ is injective; it is **transitive** if this map is surjective. For $u \in U$ the set

$$A(u) = \{a \in A : d(a) = r(a) = u\}$$

is a group with identity element u, called the **isotropy** group at u. Thus a groupoid is principal if and only if its isotropy groups are all trivial. If $a \in A$ has $u = d(a)$ and $v = r(a)$, then $x \to xax^{-1}$ is an isomorphism of $A(u)$

with $A(v)$; hence transitivity amounts to the property that the isotropy groups are conjugate to one another.

3.13.2

Here are some examples. Any group G becomes a groupoid by regarding its elements as arrows from e, the identity element, to e. So does any disjoint union of groups; in fact a disjoint union of groupoids is a groupoid in an obvious way.

An intuitively-appealing example is obtain from the continuous paths $\gamma : [0, 1] \to X$ in a topological space X. Take A to be the set of homotopy classes of such paths. Here the homotopies are required to fix the end-points, so that d and r can be unambiguously defined on a path class as the initial and final points respectively. With U the constant paths ($= X$) and the operations as in the fundamental group, axioms i–iv are satisfied and the resulting groupoid is called the **fundamental groupoid** of X. When X is path-connected it is transitive and the isotropy groups are isomorphic to the fundamental group of X.

If a group G acts as transformations of a set S, then we get a groupoid by viewing each element (x, s) of $X \times S$ as an arrow from xs to x. In other words,

$$A = X \times G, \qquad U = X (= X \times \{e\})$$

$$d(x, s) = xs, \; r(x, s) = s$$

$$(x, s)(y, t) = (x, st) \text{ if } y = xs$$

$$(x, s)^{-1} = (xs, s^{-1})$$

defines a groupoid. This groupoid is principal if and only if G acts freely.

Any equivalence relation E on a set X gives rise to a groupoid by regarding each $(x, y) \in E$ as an arrow from y to x. Thus $A = E$, $U = X$ (identified as the diagonal of $X \times X$), and the product $(x, y)(z, w)$ has the value (x, w) if $(y, z) \in E$. This is a principal groupoid, and conversely, any principal groupoid arises in this way (from the equivalence relation on U obtained by identifying u and v if there is an arrow connecting them).

3.13.3

Next we indicate how, in many cases, a C^*-algebra $C^*(\mathcal{G})$ may be attached to a groupoid $\mathcal{G} = (A, U)$. The basic idea is to use the formulas

$$(f * g)(a) = \sum_{a=bc} f(b)g(c)$$

$$f^*(a) = \overline{f(a^{-1})}$$

to give a suitable space of complex-valued functions on A the structure of a $*$-algebra, and to use its representations (i.e., $*$-homomorphisms into $\mathcal{B}(\mathcal{H})$) to construct a C^*-norm. The completion is then the groupoid C^*- algebra $C^*(\mathcal{G})$. To simplify matters, we will look first at the case of finite groupoids so that the convolution product is defined for all f and g. For the groupoid associated to a finite group G or dynamical system (X, G) it is easy to see that the groupoid C^*-algebra is precisely the group C^*-algebra $C^*(G)$ of 3.6 or the transformation group C^*-algebra of 3.8. This is clear in the group case, but in the other case it should be noted that considering (x, s) as an arrow from x to xs, rather than the other way round, would have produced the opposite algebra.

The groupoid coming from an equivalence relation is the disjoint union of the sets $F \times F$ as F runs over the equivalence classes, and consequently its C^*-algebra is the direct sum of the C^*-algebras $C^*(F \times F)$. It is easy to see that this algebra is $M_n(\mathbf{C})$. In any groupoid the characteristic functions δ_a of the singleton subsets satisfy $\delta_a * \delta_b = \delta_{ab}$ and $\delta_a^* = \delta_{a^{-1}}$; in particular if $F = \{1, 2\}$ then $\delta_{(1,1)}$, $\delta_{(1,2)}$, $\delta_{(2,1)}$, and $\delta_{(2,2)}$ is a set of 2×2 matrix units. It follows that the groupoid C^*-algebras for equivalence relations on finite sets are exactly the finite-dimensional C^*-algebras.

3.13.4

A **topological groupoid** is a groupoid $\mathcal{G} = (A, U)$ with a topology on A that makes the operations continuous. It is usually assumed that the topology is locally compact; then one wants to equip the space $C_c(\mathcal{G})$ of continuous compactly supported functions $A \to \mathbf{C}$ with the structure of a $*$-algebra and a C^*-norm. In general, this is not possible or, when possible, not unique.

When the groupoid is discrete, the operations of the preceding section are defined on $C_c(\mathcal{G})$. This algebra is then given a C^*-norm by considering its representations, in the way described in 3.8 for the crossed product, and the groupoid C^*-algebra is the completion. For example the groupoid C^*-algebra of the minimal equivalence relation (the diagonal) on a countably infinite set is c_0; the maximal equivalence relation produces $\mathcal{K}$. In fact all AF-algebras arise from discrete groupoids.

The same ideas work in the more general case of an ***r*-discrete groupoid**, a locally compact groupoid in which the subsets $r^{-1}(u)$ are discrete for all units u. For $f, g \in C_c(\mathcal{G})$, the sum defining $f * g$ in the preceding section then is over a discrete subset of a compact set, hence over a finite set. A great many interesting C^*-algebras come from such groupoids, for example, the Toeplitz algebra and the Cuntz algebras. Here is a way to get the Toeplitz algebra. Let $\mathbf{N}^+ = \mathbf{N} \cup \{\infty\}$ be the one-point compactification of $\mathbf{N} = \{0, 1, \ldots\}$, and take as arrows the set

$$A = \{(k, n) \in \mathbf{Z} \times \mathbf{N}^+ : k + n \geq 0\}$$

where (k, n) is regarded as going from n to $k + n$. Then the operations are given by

$$(k, n)(l, m) = (k + l, m) \qquad \text{provided that } l + m = n$$

$$(k, n)^{-1} = (-k, k + n)$$

and the unit space is $U = \{0\} \times \mathbf{N}^+$. This set is compact and open so that its characteristic function is in $C_c(\mathcal{G})$; in fact it is the identity element. Let $s \in C_c(\mathcal{G})$ be the characteristic function of the compact open set $\{1\} \times \mathbf{N}^+$. A short computation reveals that $s^* * s = \chi_U = 1$ and that $s * s^*$ is the characteristic function of $U \backslash \{(0, 0)\}$, so s is a proper isometry.

3.14 HILBERT C^*-MODULES

In 3.4 we considered the notion of a hermitian vector bundle E over a compact Hausdorff space X. Let $\Gamma(E)$ be the (vector) space of continuous maps $s : X \to E$ which are sections of E, in the sense that $s(x) \in E_x$ for all $x \in X$. Then $\Gamma(E)$ is in fact a module over $C(X)$, and moreover it has a $C(X)$-valued "inner product"

$$\Gamma(E) \times \Gamma(E) \longrightarrow C(X)$$

defined by $\langle s, t\rangle(x) = \langle s(x), t(x)\rangle_x$ via the given hermitian metric

$$E_x \times E_x \to \mathbf{C}$$

Thus $\Gamma(E)$ has the structure of a "Hilbert space" in which the role of the scalar field $\mathbf{C}$ is played by the C^*-algebra $C(X)$.

Extending this idea to arbitrary C^*-algebras yields the concept of a **Hilbert C^*-module**, which may usefully be thought of as a sort of noncommutative vector bundle. It has found significant applications in representation theory, KK-theory and quantum groups. Here we will give a brief introduction to this subject; for more information the reader may consult [Lan95] and [Weg93, ch.15].

3.14.1

Let A be a C^*-algebra. A **pre-Hilbert A-module** is a right A-module E together with a map

$$\langle\, ,\rangle : E \times E \longrightarrow A$$

which is A-linear (i.e., a module map) and in addition satisfies

i. $\langle x, y\rangle^* = \langle y, x\rangle$ for all $x, y \in E$;
ii. $\langle x, x\rangle \geq 0$ for all $x \in E$, with equality only for $x = 0$.

Thus we have the identities

$$\langle x, ya\rangle = \langle x, y\rangle a \text{ and } \langle xa, y\rangle = a^*\langle x, y\rangle$$

for all $x, y \in E$ and $a \in A$; in particular the A-valued inner product is linear in the second variable and conjugate–linear in the first. (Note that when A is nonunital it is necessary to assume expicitly that E is a complex linear space compatibly with the A-module structure.)

3.14.2 Lemma

For any pre-Hilbert A-module E, $||x|| = \sqrt{||\langle x, x\rangle||_A}$ *defines a norm on E.*

The main point is the triangle inequality, which follows as usual from the Cauchy–Schwarz inequality

$$||\langle x, y\rangle||_A \leq ||x|| \cdot ||y||$$

on E. This in turn results from

3.14.3 Lemma

For any $x, y \in E$, *a pre-Hilbert A-module,* $\langle x, y\rangle^*\langle x, y\rangle \leq ||x||^2\langle y, y\rangle$.

Proof

It can be assumed that $||x|| = 1$. Take $a = \langle x, y\rangle$ and compute

$$\begin{aligned} 0 &\le \langle y - xa, y - xa\rangle \\ &= \langle y, y\rangle - \langle y, x\rangle a - a^*\langle x, y\rangle + a^*\langle x, x\rangle a \\ &\le \langle y, y\rangle - \langle y, x\rangle a - a^*\langle x, y\rangle + a^* a \\ &= \langle y, y\rangle - a^* a \end{aligned}$$

making use of 2.4.2.

These arguments are valid even for A-valued *semi*-inner products. It follows that the vectors of zero norm form a submodule and that the quotient is a pre-Hilbert A-module.

3.14.4

A **Hilbert A-module** is a pre-Hilbert A-module which is complete in the norm of 3.14.2. As one might expect, the A-module structure of a pre-Hilbert module extends by continuity to the completion and one gets a Hilbert A-module.

Here are some examples. Let E be a right ideal of A (for example A itself) and take $\langle x, y\rangle = x^*y$. This is a pre-Hilbert A-module, complete if E is closed in A. More generally, any closed submodule of a Hilbert A-module is a Hilbert A-module.

For any family E_i, $i \in I$, of Hilbert A-modules, let $\sum \oplus E_i$ consist of all $x = (x_i)$ such that $\sum\langle x_i, x_i\rangle$ converges in A. Then $\langle x, y\rangle = \sum\langle x_i, y_i\rangle$ converges for all $x, y \in \sum \oplus E_i$ (by the usual argument that ℓ^2 is a Hilbert space) and defines a Hilbert A-module. In particular

$$A^n = A \oplus A \oplus \cdots \oplus A \quad \text{and} \quad \mathcal{H}_A = A \oplus A \oplus \cdots$$

are Hilbert A-modules. The latter can also be viewed as the completion of the algebraic tensor product $\ell^2 \otimes A$, where the module action and inner product are given by

$$(\xi \otimes a)b = \xi \otimes ab \quad \text{and} \quad \langle \xi \otimes a, \eta \otimes b\rangle = \langle \xi, \eta\rangle a^* b$$

It plays a distinguished role in the theory:

3.14.5 Kasparov Stabilization Theorem [Kas80]

If E is any countably-generated Hilbert A-module then $E \oplus \mathcal{H}_A \cong \mathcal{H}_A$.

In particular all countably-generated Hilbert A-modules appear as submodules—even direct summands—of $\mathcal{H}_A$.

Remark

Caution must be exercised in using the analogy between Hilbert spaces and Hilbert C^*-modules. Thus for example it is not true that submodules are necessarily complemented. If F is a submodule of a Hilbert A-module E, then the "orthogonal complement" of F, defined by

$$F^{\perp} = \{x \in E : \langle x, y \rangle = 0 \text{ for all } y \in F\}$$

satisfies $F \cap F^{\perp} = \{0\}$ but not necessarily $F + F^{\perp} = E$. For example take $E = A = C[0, 1]$ and $F = \{y : y(0) = 0\}$.

3.14.6

Let E and F be Hilbert A-modules and consider bounded complex linear operators $T : E \to F$. It is not true in general that adjoints exist, in the sense that there is a bounded complex linear $S : F \to E$ such that

$$\langle Tx, y \rangle_F = \langle x, Sy \rangle_E \text{ for all } x \in E, y \in F$$

Take $A = C[0, 1]$, $J = \{x \in A : x(0) = 0\}$ and $E = A \oplus J$. Then the operator T defined on E by $T(f, g) = (g, 0)$ has no adjoint (consider $S(1, 0)$).

If $T : E \to F$ has an adjoint then it is unique and T is said to be **adjointable**. It is not difficult to see that adjointable operators are in fact module maps, and that the space $\mathcal{L}(E)$ of adjointable operators on E is a C^*-algebra.

Adjointable operators exist, for example the analogues $\Theta_{x,y} : F \to E$ of the rank one Hilbert space operators, defined for $x \in E$ and $y \in F$ by

$$\Theta_{x,y}(z) = x\langle y, z \rangle \text{ for all } z \in F$$

It is straight-forward to verify that $(\Theta_{x,y})^* = \Theta_{y,x}$ and, if T is adjointable, that $T\Theta_{x,y} = \Theta_{Tx,y}$. In particular the C^*-subalgebra of $\mathcal{L}(E)$ generated by the $\Theta_{x,y}$, denoted $\mathcal{K}(E)$ by analogy with the compact operators, is an ideal of $\mathcal{L}(E)$.

Here are some examples. If A is unital then the morphism $A \to \mathcal{L}(E)$ given by $a \to L_a$, left multiplication by a, is an isomorphism (since for $T \in \mathcal{L}(A)$ and $a \in A$ we have $Ta = (T1)a$). Moreover $L_a = \Theta_{a,1}$ and so $\mathcal{K}(A) = \mathcal{L}(A) \cong A$.

For any Hilbert A-module E, $\mathcal{K}(E^n) \cong M_n(\mathcal{K}(E))$. In particular, for A unital, $\mathcal{K}(A^n) \cong M_n(A)$. It follows that $\mathcal{K}(\mathcal{H}_A) \cong A \otimes \mathcal{K}$, where $\mathcal{K}$ is the algebra of compact operators (see 3.10.2).

3.14.7 Theorem [Kas80]

For any Hilbert C^*-module E, the multiplier algebra $M(\mathcal{K}(E))$ is isomorphic to $\mathcal{L}(E)$.

Multiplier algebras were introduced in 3.12, where in particular it was shown that the multiplier algebra of $\mathcal{K}(\mathcal{H})$ is $\mathcal{B}(\mathcal{H})$. The theorem is proved by showing that the desired isomorphism is given by $T \to (T_1, T_2)$, where

$$T_1(\Theta_{x,y}) = \Theta_{Tx,y} \text{ and } T_2(\Theta_{x,y}) = \Theta_{x,T^*y}$$

extend to $\mathcal{K}(E)$ by linearity and continuity.

For example when $A = C(X)$ we have $\mathcal{K}(\mathcal{H}_A) \cong A \otimes \mathcal{K} \cong C(X, \mathcal{K})$ and

$$\mathcal{L}(\mathcal{H}_A) \cong M(C(X, \mathcal{K})) \cong C_{*s}(X, \mathcal{B}(\mathcal{H}))$$

the algebra of bounded and $*$-strongly continuous functions from X to $\mathcal{B}(\mathcal{H})$. We recall that $f : X \to \mathcal{B}(\mathcal{H})$ is **$*$-strongly continuous** if both f and f^* are strongly continuous, where f^* is defined by $f^*(x) = (f(x))^*$.

Chapter Four

Von Neumann Algebras

4.1 INTRODUCTION

Von Neumann algebras were introduced by F. J. Murray and John von Neumann in the first [MvN36] of the famous series of papers *On Rings of Operators*. They were the first class of operator algebras (or C^*-algebras) to be introduced and studied systematically, but it should be noted that their behavior differs strikingly in several important respects from that of arbitrary C^*-algebras. Just as C^*-algebras can usefully be thought of as noncommutative topological spaces, so von Neumann algebras can be regarded as noncommutative measure spaces. Thus, in particular, while a C^*-algebra may contain no nontrivial projections (i.e., open-and-closed sets), any von Neumann algebra contains an abundance of them (measurable sets). Another difference is that infinite-dimensional von Neumann algebras are necessarily nonseparable, whereas the interesting examples of infinite-dimensional C^*-algebras are usually separable. For example, $C[0, 1]$ is separable but $L^\infty[0, 1]$ is not.

Murray and von Neumann gave several reasons for studying "rings of operators." They arise naturally in attempts to extend the theory of unitary representations of groups beyond the classical case, as well as in the quantum mechanical formalism based on Hilbert spaces and operators. And they are objects of great intrinsic interest: They are infinite-dimensional and yet tractable, and they present fascinating and fundamental new phenomena such as continuous dimensionality.

The original Murray-von Neumann series is [MvN36], [MvN37], [Neu40] and [MvN43]. Good general references for the subject include [Dix81] (the first edition of this classic was published in 1957), [Sak71], [Tak79], [StZ79], [Sun87] and [KaR86]; [Con94] has a masterful overview.

4.1.1

The **weak** and **strong operator topologies** on $\mathcal{B}(\mathcal{H})$ are defined as follows: For any net (T_α) in $\mathcal{B}(\mathcal{H})$, $T_\alpha \to T$ weakly (strongly) if $T_\alpha \xi \to T\xi$ weakly (in norm) for all $\xi \in \mathcal{H}$.

4.1.2

Obviously $T_\alpha \to T$ in norm implies that $T_\alpha \to T$ strongly, which in turn implies that $T_\alpha \to T$ weakly. These implications are strict, as the following examples show.

Example 1. If $\{P_\lambda : \lambda \in \Lambda\}$ is an infinite set of pairwise orthogonal projections in $\mathcal{B}(\mathcal{H})$, then the net $(P_F : F \subset \Lambda,\ F$ finite), where P_F denotes the sum of the P_λ for $\lambda \in F$, converges strongly to the projection P on the closed linear span in $\mathcal{H}$ of the ranges $P_\lambda \mathcal{H}$. But $\|P - P_F\| = 1$ for all F so (P_F) does not converge to P in norm.

Example 2. If $\{e_n : n \in \mathbf{N}\}$ is an orthonormal basis of $\mathcal{H}$ and $U_n = e_1 \otimes e_n$ (i.e., $U_n \xi = (\xi, e_n)e_1$), then $U_n \to 0$ strongly. Since $U_n^* \xi = (\xi, e_1)e_n$, we see that $U_n^* \not\to 0$ strongly and yet $U_n^* \to 0$ weakly.

4.1.3

Evidently $T_n \to T$ weakly implies that $T_n^* \to T^*$ weakly; in other words, the adjoint operation is continuous in the weak operator topology. As Example 2 shows this is false for the strong operator topology.

4.2 THE DOUBLE COMMUTANT THEOREM

The **commutant** of a subset $\mathcal{S}$ of $\mathcal{B}(\mathcal{H})$ is defined by

$$\mathcal{S}' = \{T \in \mathcal{B}(\mathcal{H}) : TS = ST \text{ for all } S \in \mathcal{S}\}$$

It is a unital subalgebra of $\mathcal{B}(\mathcal{H})$, closed in the weak operator topology, hence in the strong operator topology. The **double commutant** of $\mathcal{S}$ is $\mathcal{S}'' = (\mathcal{S}')'$.

In the next theorem Alg $\mathcal{S}$ denotes the subalgebra generated by $\mathcal{S}$ in the strictly algebraic sense: Its elements are the finite linear combinations of the finite products of elements of $\mathcal{S}$. We say that $\mathcal{S}$ is **self-adjoint** if $S^* \in \mathcal{S}$ whenever $S \in \mathcal{S}$ and that $\mathcal{S}$ is **nondegenerate**, or that it **acts nondenegerately**, if the only $\xi \in \mathcal{H}$ for which $S\xi = 0$ for all $S \in \mathcal{S}$ is $\xi = 0$. The proof makes use of the **n-fold inflation** $\mathcal{S}^{(n)}$, which consists of all operators $S \oplus \cdots \oplus S$, $S \in \mathcal{S}$, acting on $\mathcal{H}^n = \mathcal{H} \oplus \cdots \oplus \mathcal{H}$.

4.2.1 Theorem

Let $\mathcal{S} \subset \mathcal{B}(\mathcal{H})$ be self-adjoint and nondegenerate. Then $\operatorname{Alg} \mathcal{S}$ *is strongly (hence weakly) dense in $\mathcal{S}''$.*

Proof

It is easy to see that $\operatorname{Alg} \mathcal{S} \subset \mathcal{S}''$, so consider $T \in \mathcal{S}''$. To construct a net in $\operatorname{Alg} \mathcal{S}$ converging strongly to T, it is enough to show that for any finite set of vectors $\xi_1, \ldots, \xi_n$ and $\varepsilon > 0$, there is $S \in \operatorname{Alg} \mathcal{S}$ such that $\|T\xi_i - S\xi_i\| < \varepsilon$ for $i = 1, 2, \ldots, n$. Consider the case of one vector ξ_0, and let $\mathcal{M}$ be the closure of $(\operatorname{Alg} \mathcal{S})\xi_0$. What we have to show is precisely that $T\xi_0 \in \mathcal{M}$. Now $S\mathcal{M} \subset \mathcal{M}$ for every $S \in \mathcal{S}$, and so $PSP = SP$, where P is the projection on $\mathcal{M}$. Since $\mathcal{S}$ is self-adjoint, it follows that $PS^*P = S^*P$; hence $P \in \mathcal{S}'$. But $T \in \mathcal{S}''$, so $TP = PT$ and $T\mathcal{M} \subset \mathcal{M}$. It will therefore be enough to see that $\xi_0 \in \mathcal{M}$. This comes from the nondegeneracy:

$$S(I - P)\xi_0 = (I - P)S\xi_0 = (I - P)PS\xi_0 = 0$$

for all $S \in \mathcal{S}$, so $(I - P)\xi_0 = 0$.

The general case, for vectors $\xi_1, \ldots, \xi_n$, is reduced to the case just handled by taking $\xi_0 = \xi_1 \oplus \cdots \oplus \xi_n \in \mathcal{H}^n$ and replacing $\mathcal{S}$ by the inflation $\mathcal{S}^{(n)} \subset \mathcal{B}(\mathcal{H}^n)$.

4.2.2 ***Corollary*** (Von Neumann's Double Commutant Theorem)

For any unital $*$-subalgebra $\mathcal{A} \subset \mathcal{B}(\mathcal{H})$, $\mathcal{A}''$ coincides with the strong and weak operator closures of $\mathcal{A}$.

4.2.3

Let $\mathcal{A}$ be a $*$-subalgebra of $\mathcal{B}(\mathcal{H})$, let

$$\mathcal{N} = \{\xi \in \mathcal{H} : A\xi = 0 \text{ for all } A \in \mathcal{A}\}$$

and let $\mathcal{A}_e = \mathcal{A}|\mathcal{N}^\perp$. Then $\mathcal{A} = \mathcal{A}_e \oplus \{0\}$, $\mathcal{A}' = \mathcal{A}_e' \oplus \mathcal{B}(\mathcal{N})$, and $\mathcal{A}'' = \mathcal{A}_e'' \oplus \mathbf{C}I_{\mathcal{N}}$, but the strong and weak operator closures of $\mathcal{A}$ are $\mathcal{A}_e'' \oplus \{0\}$.

4.3 DEFINITION AND EXAMPLES

A **von Neumann algebra** is a unital $*$-subalgebra $\mathcal{R}$ of $\mathcal{B}(\mathcal{H})$ that satisfies one (hence all) of $\mathcal{R} = \mathcal{R}''$, $\mathcal{R}$ is weakly closed, $\mathcal{R}$ is strongly closed.

Of course a von Neumann algebra is in particular a C^*-algebra. An important difference between the two notions is that certain basic constructions that may lead outside of a C^*-algebra can be carried out within any von Neumann algebra. For example, any von Neumann algebra contains the spectral projections of its normal elements (this is clear from the double commutant characterization).

4.3.1

It is not immediately obvious what the appropriate notion of morphism is for von Neumann algebras. This is related to the fact that the definition of these algebras is not intrinsic; we will return to this later (4.4). Von Neumann algebras $\mathcal{R}_1$ and $\mathcal{R}_2$ on Hilbert spaces $\mathcal{H}_1$ and $\mathcal{H}_2$ are **spatially isomorphic** if there is a unitary $U : \mathcal{H}_1 \to \mathcal{H}_2$ that carries $\mathcal{R}_1$ to $\mathcal{R}_2$ (i.e., $U\mathcal{R}_1U^* = \mathcal{R}_2$). They are **$*$-isomorphic** if they are isomorphic as C^*-algebras. The mapping $T \to T \oplus \cdots \oplus T$ from $\mathcal{B}(\mathcal{H})$ to $\mathcal{B}(\mathcal{H}^n)$ is an example of a $*$-isomorphism that is not spatial.

4.3.2

For $\mathcal{S}, \mathcal{T} \subset \mathcal{B}(\mathcal{H})$ observe that $\mathcal{S} \subset \mathcal{S}''$ and that $\mathcal{S} \subset \mathcal{T}$ implies that $\mathcal{S}' \supset \mathcal{T}'$. It follows that always $\mathcal{S}' = \mathcal{S}'''$; hence the commutant $\mathcal{S}'$ of any self-adjoint set of operators $\mathcal{S} \subset \mathcal{B}(\mathcal{H})$ is a von Neumann algebra.

Here is a situation in which this observation is of interest. Let G be a group and $\pi : G \to \mathcal{B}(\mathcal{H})$ a unitary representation of G (see 3.6.2). Projections P in the commutant $\pi(G)'$ correspond to **subrepresentations** $\pi|_P$ of π, obtained by restricting the operators $\pi(g)$ to the invariant subspace $P\mathcal{H}$. Moreover it is easy to see that projections P and Q in $\pi(G)'$ are Murray-von Neumann equivalent (3.10.2) there if and only if the corresponding subrepresentations are unitarily equivalent (this means that there is a unitary operator $U : P\mathcal{H} \to Q\mathcal{H}$ such that $U\pi(g) = \pi(g)U$ on $P\mathcal{H}$ for all $g \in G$). See [Mac76].

4.3.3

If $\mathcal{H} = \mathbf{C}^n$ is finite-dimensional, then all three operator topologies on $\mathcal{B}(\mathcal{H}) = M_n(\mathbf{C})$ coincide and any $*$-subalgebra is a C^*-algebra and (if it contains the identity operator $I_{\mathcal{H}}$) a von Neumann algebra. As mentioned in 3.3, any such algebra is $*$-isomorphic to a finite direct sum of full matrix algebras over $\mathbf{C}$. Note, however, that it may not be possible to take this isomorphism to be spatial. In fact the underlying spaces need not be isomorphic at all. For example, the algebra of all operators of the form $\begin{pmatrix} aI & bI \\ cI & dI \end{pmatrix}$ on $\mathbf{C}^6$ (where I is the identity operator on $\mathbf{C}^3$) is isomorphic to the algebra $M_2(\mathbf{C})$ on $\mathbf{C}^2$. The commutant of this algebra consists of all operators $\begin{pmatrix} A & 0 \\ 0 & A \end{pmatrix}$ for $A \in M_3(\mathbf{C})$; it is isomorphic to $M_3(\mathbf{C})$.

4.3.4

For $\phi \in L^\infty[a, b]$ let M_ϕ be the corresponding multiplication operator on $\mathcal{H} = L^2[a, b]$, defined by $M_\phi f = \phi f$ for $f \in L^2[a, b]$. The map $\phi \to M_\phi$ carries L^∞ $*$-isomorphically to a subalgebra of $\mathcal{B}(L^2[a, b])$, and it is customary to identify L^∞ with its image. Viewed in this way, L^∞ is a von Neumann algebra. Since it is commutative, we have $(L^\infty)' \supset L^\infty$, and in fact this is an equality. More generally, for any σ-finite measure space the range of

$$L^\infty(X, \mathcal{M}, \mu) \to \mathcal{B}(L^2(X, \mathcal{M}, \mu))$$

is equal to its own commutant [Dou72, 4.22] and hence is a von Neumann algebra. It is interesting to note that the weak operator topology on $L^\infty(\mu)$ is precisely its weak*-topology as the dual of $L^1(\mu)$. This comes easily from the fact that $L^1(\mu)$ consists of all products $\xi\overline{\eta}$ with ξ and η in $L^2(\mu)$ [Dou72, 4.51].

4.3.5

From a measure space $(X, \mathcal{M}, \mu)$ and a separable Hilbert space $\mathcal{H}$, one can construct a Hilbert space $L^2(X, \mathcal{M}, \mu; \mathcal{H})$ of vector-valued functions as follows: Call $\xi : X \to \mathcal{H}$ **weakly measurable** if $x \to \langle \xi(x), \eta \rangle$ is measurable for all $\eta \in \mathcal{H}$. Then $x \to \|\xi(x)\|^2$ is measurable (this uses the separability of $\mathcal{H}$), and the Hilbert space in question consists of those ξ for which this function is integrable, modulo null functions. Let $\mathcal{R}$ be a von Neumann algebra on $\mathcal{H}$, and denote by $L^\infty(X, \mathcal{M}, \mu; \mathcal{R})$ the bounded functions $F : X \to \mathcal{R}$ (modulo null functions) such that $x \to \langle F(x)\xi, \eta \rangle$ is measurable for all $\xi, \eta \in \mathcal{H}$. As in 4.3.3 the image of the obvious map

$$L^\infty(X, \mathcal{M}, \mu; \mathcal{R}) \to \mathcal{B}(L^2(X, \mathcal{M}, \mu; \mathcal{H}))$$

is a von Neumann algebra; its commutant is the image of $L^\infty(X, \mathcal{M}, \mu; \mathcal{R}')$.

4.3.6

By the **direct sum** $\sum \oplus \mathcal{R}_\lambda$ of a family $\{\mathcal{R}_\lambda\}$ of von Neumann algebras is meant the ℓ^∞-direct sum (defined in 3.3). If $\mathcal{R}_\lambda$ acts on $\mathcal{H}_\lambda$, then $\sum \oplus \mathcal{R}_\lambda$ is a von Neumann algebra on $\sum \oplus \mathcal{H}_\lambda$ with commutant $\sum \oplus \mathcal{R}'_\lambda$.

4.3.7

An abelian (= commutative) subalgebra of $\mathcal{B}(\mathcal{H})$ is said to be **maximal abelian** if it is not properly contained in any abelian subalgebra of $\mathcal{B}(\mathcal{H})$. Any abelian subalgebra $\mathcal{A}$ satisfies $\mathcal{A} \subset \mathcal{A}'$, and it is true (and almost obvious) that equality is equivalent to maximality. Thus the L^∞ algebras considered in 4.3.4 are maximal abelian.

The self-adjoint maximal abelian subalgebras, known as **masas**, are von Neumann algebras and play an important role in the structure theory. For separable spaces a surprisingly strong statement is valid: Any masa on a separable Hilbert space is spatially isomorphic to exactly one of $L^\infty[0, 1]$, $\ell^\infty(n)$, or $L^\infty[0, 1] \oplus \ell^\infty(n)$. Here $1 \le n \le \aleph_0$, and $\ell^\infty(n)$ denotes the space of bounded complex sequences of length n [KaR86, 9.4.1]. It should also be noted that any direct sum of masas is a masa (consider the commutant).

4.3.8

The structure of abelian von Neumann algebras can be described rather precisely in terms of inflations of masas. The n-fold inflation was introduced at the beginning of 4.2; here we require it for arbitrary cardinal numbers n. A proof of the following theorem can be found in [KaR86, 9.4].

Theorem

If $\mathcal{A}$ is an abelian von Neumann algebra on Hilbert space $\mathcal{H}$, then for each $n \le \dim \mathcal{H}$ there is a masa $\mathcal{Z}_n$ such that $\mathcal{A}$ is spatially isomorphic to $\sum \oplus \mathcal{Z}_n^{(n)}$. Moreover the $\mathcal{Z}_n$'s are unique up to spatial isomorphism. If $\mathcal{Z}_n$ acts on $\mathcal{K}_n$, then the restriction of $\mathcal{A}$ to the subspace corresponding to the direct sum of one copy of each $\mathcal{K}_n$ is an injection, and hence $\mathcal{A}$ is $*$-isomorphic to the masa $\sum \oplus \mathcal{Z}_n$.

This result is closely related to the classification of self-adjoint operators, up to unitary equivalence, via measures and spectral multiplicity. For a self-adjoint operator H on a separable Hilbert space, this asserts that there is a finite

Borel measure μ on $\sigma(H)$ and a Borel decomposition

$$\sigma(H) = \sigma_1 \dot{\cup} \sigma_2 \dot{\cup} \cdots \dot{\cup} \sigma_\infty$$

such that $H \cong \sum \oplus H_n$, where H_n is the n-fold inflation of M_t (multiplication by the independent variable) on $L^2(\mu|\sigma_n)$. Moreover the measure class of μ is uniquely determined, as are the σ_n modulo sets of measure zero.

To see the connection, apply the theorem to the von Neumann algebra $\{H\}''$ generated by H to get $\{H\}'' \cong \sum \oplus \mathcal{Z}_n^{(n)}$. Let $I = \sum E_n$ be the corresponding decomposition into projections in $\{H\}''$, and let H_n be the restriction of H to the range of E_n, so that $H_n \cong K_n^{(n)} \in \mathcal{Z}_n^{(n)}$. Then one takes $\sigma_n = \sigma(H_n)$ and $\mu = \sum \mu_n$, the μ_n being obtained by applying the following result to the masa $\mathcal{Z}_n$ and its generator K_n.

4.3.9 Theorem

i. *Any masa on a separable space has a cyclic vector.*

ii. *If H is a cyclic self-adjoint operator on a separable space, then there exists a finite positive Borel measure μ on $\sigma(H)$ such that the functional calculus $C(\sigma(H)) \to C^*(H)$ of 2.2.5 extends uniquely to a spatial isomorphism $L^\infty(\mu) \to \{H\}''$.*

Proof

Let $\mathcal{Z}$ be a masa on a separable space $\mathcal{H}$, and let $\{\xi_n\}$ be a maximal set of unit vectors such that the spaces $\mathcal{Z}\xi_n$ are pairwise orthogonal. By separability the set is countable; let $\xi = \sum \frac{1}{n}\xi_n$. By maximality of $\mathcal{Z}$ the orthogonal projection on $(\mathcal{Z}\xi_n)^-$ is in $\mathcal{Z}$ for all n. It follows that $\mathcal{Z}\xi$ is dense in $\mathcal{H}$: If $\eta \perp \mathcal{Z}\xi$, then $\eta \perp \mathcal{Z}\xi_n$ for all n, and so $\eta = 0$ by maximality of $\{\xi_n\}$.

Let ξ_0 be cyclic for H. Then

$$f \to \langle f(H)\xi_0, \xi_0\rangle, \qquad f \in C(\sigma(H))$$

is positive and linear, so by the Riesz representation theorem there is a positive Borel measure μ on $\sigma(H)$ such that

$$(f(H)\xi_0, \xi_0) = \int f d\mu \qquad \text{for all } f \in C(\sigma(H))$$

Define a linear map U : polynomials $\to \mathcal{H}$ by $Up = p(H)\xi_0$. Then

$$\langle Up, Uq\rangle = \langle q(H)^* p(H)\xi_0, \xi_0\rangle = \int \bar{q}p\,d\mu = \langle p, q\rangle_{L^2(\mu)}$$

so U extends to an isometry $L^2(\mu) \to \mathcal{H}$. The range of U contains ξ_0, $H\xi_0$, $H^2\xi_0, \cdots$, and so U is unitary. To see that U respects the functional calculus, it is enough (by the Stone-Weierstrass theorem) to show that for any polynomials p we have $UM_pU^* = p(H)$, where M_p denotes multiplication by p on $L^2(\mu)$. But for any polynomial q

$$(UM_p)q = U(M_pq) = U(pq) = p(H)q(H)\xi_0 = p(H)Uq$$

as required. In particular, it follows that $UL^\infty(\mu)U^* \supset \{H\}''$. The other inclusion results from the facts that $C(\sigma(H))$ is weak*-dense in $L^\infty(\mu)$ and that on $L^\infty(\mu)$ the weak* and weak operator topologies coincide (see 4.3.4).

4.3.10

The tensor product of von Neumann algebras is defined as follows: If $\mathcal{R}_1$ and $\mathcal{R}_2$ are von Neumann algebras on $\mathcal{H}_1$ and $\mathcal{H}_2$, then $\mathcal{R}_1 \mathbin{\overline{\otimes}} \mathcal{R}_2$ is the weak operator closure of the algebraic tensor product $\mathcal{R}_1 \otimes \mathcal{R}_2$ viewed as operators on the tensor product Hilbert space $\mathcal{H}_1 \otimes \mathcal{H}_2$ (see 3.7.4). As in the case of Hilbert spaces, we will omit the bar and use the same notation to denote both the algebraic and von Neumann algebra tensor products of von Neumann algebras; the context will usually make it clear which is intended.

Here are some examples. The algebra $L^\infty(X, \mathcal{M}, \mu; \mathcal{R})$ of 4.3.5 is spatially isomorphic to $L^\infty(\mu) \otimes \mathcal{R}$. The n-fold inflation $\mathcal{R}^{(n)}$ of an algebra $\mathcal{R}$, described in 4.2, is spatially isomorphic to $\mathcal{R} \otimes \mathbf{C}I_n$ (where I_n is the identity operator on an n-dimensional space). It is true in general that $(\mathcal{R}_1 \otimes \mathcal{R}_2)' = \mathcal{R}_1' \otimes \mathcal{R}_2'$ [KaR86, ch. 11]. Thus $(L^\infty(\mu) \otimes \mathcal{R})' = L^\infty(\mu) \otimes \mathcal{R}'$ and

$$\mathcal{R}^{(n)\prime} \cong (\mathcal{R} \otimes \mathbf{C}I_n)' = \mathcal{R}' \otimes \mathcal{B}(\mathcal{H}_n)$$

where $\dim \mathcal{H}_n = n$.

4.3.11

The **center** $\mathcal{Z}(\mathcal{R})$ of a ring $\mathcal{R}$ consists of the elements of $\mathcal{R}$ that commute with everything in $\mathcal{R}$. For a von Neumann algebra $\mathcal{R}$ it is obvious that

$\mathcal{Z}(\mathcal{R}) = \mathcal{R} \cap \mathcal{R}'$; hence $\mathcal{Z}(\mathcal{R})$ is again a von Neumann algebra (abelian, of course). Direct sum decompositions of a von Neumann algebra correspond to to decompositions of the identity operator into pairwise orthogonal projections in the center of the algebra.

4.4 THE σ-WEAK TOPOLOGY

The **σ-weak topology** on $\mathcal{B}(\mathcal{H})$ is its weak* topology as the dual of the trace class $\mathcal{B}_1(\mathcal{H})$ [Mur90, 4.2.3]. We recall that T is **trace class** if

$$\mathrm{Tr}\,|T| = \sum \langle |T|\xi_i, \xi_i\rangle < \infty$$

for any orthonormal basis $\{\xi_i\}$ of $\mathcal{H}$, where $|T| = (T^*T)^{1/2}$, and that then

$$\mathrm{Tr}(T) = \sum \langle T\xi_i, \xi_i\rangle$$

converges to a limit independent of the orthonormal basis. Moreover $\mathcal{B}_1(\mathcal{H})$ is an ideal of $\mathcal{B}(\mathcal{H})$, and the pairing between them is given by

$$(T, S) \to \mathrm{Tr}(TS) \qquad \text{for all } T \in \mathcal{B}_1(\mathcal{H}) \text{ and } S \in \mathcal{B}(\mathcal{H})$$

In particular, σ-weak convergence implies weak operator convergence (take T of rank 1), and so any von Neumann algebra $\mathcal{R}$ on $\mathcal{H}$ is σ-weakly closed in $\mathcal{B}(\mathcal{H})$. It follows that $\mathcal{R}$ is a dual space (of $\mathcal{B}_1(\mathcal{H})$ modulo the annihilator of $\mathcal{R}$) and that the relative σ-weak topology on $\mathcal{R}$ coincides with its weak* topology. An important technical point here is that $\mathcal{R}$ is uniquely a dual space: any two preduals have the same canonical image in $\mathcal{R}^*$ [Sak71, 1.13.3]. Thus the σ-weak topology on a von Neumann algebra is intrinsic. In particular, it appears that the appropriate morphisms for von Neumann algebras are the σ-weakly continuous $*$-homomorphisms. If $\varphi : \mathcal{R}_1 \to \mathcal{R}_2$ is such a map then $\ker\varphi$ is a direct summand of $\mathcal{R}_1$ and $\mathrm{im}\,\varphi$ is σ-weakly closed [Sak71, 1.16.2]. But note that (because the predual is unique) any $*$-isomorphism of von Neumann algebras is automatically σ-weakly continuous.

Although infinite-dimensional von Neumann algebras are nonseparable, their preduals may very well be separable. In fact a von Neumann algebra has separable predual precisely when it is $*$-isomorphic to a von Neumann algebra on a separable Hilbert space.

Sakai has shown that the above dual space property of von Neumann algebras is actually a characterization: any C^*-algebra that is a dual space is

$*$-isomorphic to a von Neumann algebra [Sak71, 1.16.7]. Another characterization, as C^*-algebras with certain order-theoretical properties, was found by Kadison [KaR86, 7.6.38]. When viewed abstractly as C^*-algebras with additional properties, von Neumann algebras are frequently referred to as **W^*-algebras**.

As noted in 4.3, von Neumann algebras contain plenty of projections; several interesting partial characterizations, due initially to Rickart but developed mainly by Kaplansky, have been based on this observation [Kap68].

4.5 MURRAY-VON NEUMANN EQUIVALENCE

A fundamental concept in Murray and von Neumann's work on rings of operators is the equivalence relation on projections known as **Murray-von Neumann equivalence**. This was introduced in 3.10.2 and interpreted in the context of group representations in 4.3.2. In this section we look at some of its deeper properties, not valid for C^*-algebras in general.

4.5.1

The difference appears already in the partially ordered set $P(\mathcal{R})$ of projections in $\mathcal{R}$. For C^*-algebras this need not be a lattice, but for von Neumann algebras it is in fact a **complete lattice**: For any family $\{p_i\}$ of projections in a von Neumann algebra $\mathcal{R}$ on a Hilbert space $\mathcal{H}$, the projection p on the closed linear span $\bigvee p_i\mathcal{H}$ is in $\mathcal{R}$ and has the properties required of the supremum:

i. $p_i \leq p$ for all i.
ii. $q \in P(\mathcal{R})$ and $p_i \leq q$ for all $i \implies p \leq q$.

The infimum is given by the projection on $\bigcap p_i\mathcal{H}$. This lattice is **complemented**, in the sense that for any p there is q with $p \vee q = 1$ and $p \wedge q = 0$. In fact the map $\varphi : p \to p^\perp = 1 - p$ is an **orthocomplementation**, namely an order-reversing bijection of period 2 taking any element to a complement. It need not satisfy the **distributive** law

$$(p \vee q) \wedge r = (p \wedge r) \vee (q \wedge r)$$

and indeed this identity characterizes the abelian algebras. The weakened form obtained by restricting to projections with $p \perp q$ and $p \leq r$ is called the **orthomodular** law and is valid in general.

4.5.2

A basic example of equivalence of projections arises from the polar decomposition $x = v|x|$, $\ker v = \ker x$. If x belongs to a von Neumann algebra $\mathcal{R}$, then so do $|x|$, v, the support projection p (on $(\ker x)^\perp$), and the range projection q (on $\overline{x\mathcal{H}}$); this comes easily from the double commutant version of the definition. Moreover $v^*v = p$ and $vv^* = q$, so $p \sim q$.

This has a useful immediate consequence: if $e, f \in P(\mathcal{R})$ and $e\mathcal{R}f \neq \{0\}$, then there are nonzero subprojections $p \leq e$ and $q \leq f$ with $p \sim q$.

What is the meaning of $e\mathcal{R}f = \{0\}$? To answer this question, consider the projection z on the span $\mathcal{M}$ of the ranges of everything in $\mathcal{R}f$. It is clear that $z \in \mathcal{R}$, $f \leq z$, and $e \perp z$. But also every $x \in \mathcal{R}$ leaves $\mathcal{M}$ invariant, and so $z \in \mathcal{R}'$. Thus z belongs to the center of $\mathcal{R}$, and we have a direct sum decomposition $\mathcal{R} = z\mathcal{R} \oplus z^\perp\mathcal{R}$ with e and f in distinct summands.

4.5.3

Consider next the relation $p \preceq q$ on $P(\mathcal{R})$ defined by the requirement that $p \sim p'$ for some $p' \leq q$. It has the following fundamental properties:

i. **Cantor-Bernstein property**. If $p \preceq q$ and $q \preceq p$, then $p \sim q$.
ii. **Generalized comparability**. For any $p, q \in P(\mathcal{R})$, there exists a projection $z \in \mathcal{Z}(\mathcal{R})$ such that $pz \preceq qz$ and $qz^\perp \preceq pz^\perp$.

The Cantor-Bernstein property is named for the analogous result in set theory, and in fact it can be proved in exactly the same way. By hypothesis there are $q', p' \in P(\mathcal{R})$ with $p \sim q' \leq q$ and $q \sim p' \leq p$; hence $q' \sim p'' \leq p'$ for some $p'' \in P(\mathcal{R})$. So we have $p'' \leq p' \leq p$ and $p \sim p''$, and we have to show that these conditions imply that $p \sim p'$. Choose $u \in \mathcal{R}$ with $u^*u = p$ and $uu^* = p''$. Let $e_{2n} = u^n p u^{*n}$ and $e_{2n+1} = u^n p' u^{*n}$ for $n \geq 0$. Then $e_0 = p$, $e_1 = p'$, and $e_2 = p''$, and it follows that the sequence $\{e_n\}$ is decreasing. Let $f = \inf e_n$ and $f_n = e_n - e_{n+1}$ for $n \geq 0$ so that

$$p = \sum_{n=0}^{\infty} f_n + f \quad \text{and} \quad p' = \sum_{n=1}^{\infty} f_n + f$$

Now u takes e_n to e_{n+2} (i.e., the partial isometry ue_n has these initial and final projections), hence f_n to f_{n+2}, and therefore $\sum_{n=0}^{\infty} f_{2n} \sim \sum_{n=1}^{\infty} f_{2n}$. Adding $\sum_{n=0}^{\infty} f_{2n+1} + f$ to both sides of this gives $p \sim p'$ as required.

Generalized comparability is established by a maximality argument. Let $\{(p_i, q_i)\}$ be a maximal family of pairs of equivalent projections with the

p_i pairwise orthogonal subprojections of p and the q_i pairwise orthogonal subprojections of q; let $p = \sum p_i + e$ and $q = \sum q_i + f$. Then $\sum p_i \sim \sum q_i$ (since the sum of partial isometries with pairwise orthogonal support and range projections converges in the weak operator topology), and $e\mathcal{R}f = 0$ (by 4.5.2 and maximality). Hence, again by 4.5.2, there is a projection $z \in \mathcal{Z}(\mathcal{R})$ with $e \le z$ and $f \perp z$. Then

$$qz = \left(\sum q_i\right) z + fz = \left(\sum q_i\right) z \sim \left(\sum p_i\right) z \le pz$$

and similarly $pz^{\perp} \preceq qz^{\perp}$.

We remark that the decomposition in property ii is unique only up to a central summand in which $p \sim q$.

4.6 TYPE I ALGEBRAS

A projection p in a von Neumann algebra $\mathcal{R}$ is **abelian** if $p\mathcal{R}p$ is abelian, and $\mathcal{R}$ is **type I** if the identity element 1 can be decomposed as an orthogonal sum of abelian projections. Any projection equivalent to an abelian projection is itself abelian, and it follows (using 4.5.2) that a von Neumann algebra is type I if and only if every nonzero projection contains a nonzero abelian projection. Of course any abelian algebra $\mathcal{A}$ is type I, as is $\mathcal{B}(\mathcal{H})$; more generally so is $\mathcal{A} \otimes \mathcal{B}(\mathcal{H})$, since the projections $1 \otimes p_i$ (where the p_i's are the projections on an orthogonal basis for $\mathcal{H}$) are abelian and add up to the identity $1 \otimes 1$. The decomposition of 1 into abelian projections is by no means unique, but, as we will now see, it can be done in a sufficiently organized fashion to permit a complete description of type I algebras in terms of abelian algebras.

4.6.1

The abelian projections $1 \otimes p_i$ of the above paragraph are pairwise equivalent, and in general a type I algebra is said to be **homogeneous** if the identity operator can be decomposed into pairwise orthogonal and equivalent abelian projections. Such a decomposition is not unique, but its cardinality, say n, is [Tak79, ch.V, 1.26] and $\mathcal{R}$ is said to be ***n*-homogeneous** or of **type $\mathbf{I}_n$**. Evidently the 1-homogeneous algebras are precisely the abelian algebras.

Consider now a von Neumann algebra $\mathcal{R}$ on $\mathcal{H}$ such that the commutant $\mathcal{R}'$ happens to be abelian. If $\mathcal{R}'$ is spatially isomorphic to $\sum \oplus \mathcal{Z}_n^{(n)}$ as in 4.3.8, then by 4.3.6 and 4.3.10 we have

$$\mathcal{R} = \mathcal{R}'' \cong \sum \oplus (\mathcal{Z}_n^{(n)})' = \sum \oplus (\mathcal{Z}_n' \otimes \mathcal{B}(\mathcal{H}_n)) = \sum \oplus (\mathcal{Z}_n \otimes \mathcal{B}(\mathcal{H}_n))$$

This displays $\mathcal{R}$ as a direct sum of n-homogeneous algebras for $1 \leq n \leq \dim \mathcal{H}$ (some terms may be absent). It also shows that $\mathcal{R}$ is type I, and in fact such a decomposition exists for any type I algebra:

4.6.2 Theorem

Let $\mathcal{R}$ be a type I von Neumann algebra on a Hilbert space $\mathcal{H}$. Then there is a unique decomposition

$$\mathcal{R} = \sum \oplus \{\mathcal{R}_n : 1 \leq n \leq \dim \mathcal{H}\}$$

into n-homogeneous algebras, some of which may be absent. Moreover there are abelian algebras $\mathcal{A}_n$ and spatial isomorphisms

$$\mathcal{R}_n \cong \mathcal{A}_n \otimes \mathcal{B}(\mathcal{H}_n), \qquad 1 \leq n \leq \dim \mathcal{H}$$

where $\dim \mathcal{H}_n = n$.

Proof

The key point is that any nonzero direct summand of a type I algebra $\mathcal{R}$ necessarily contains a nonzero homogeneous summand. Assuming this for the moment, then since the center of $\mathcal{R}$ is itself a von Neumann algebra, a straightforward maximality argument will yield the first statement of the theorem.

Now given a central projection $z_0 \neq 0$ in a type I algebra, find an abelian projection $0 \neq q_1 \leq z_0$ and a maximal orthogonal family $\{q_i\}$ of projections equivalent to q_1. If $r = z_0 - \sum q_i$, then $q_1 \preceq r$ is impossible, so by generalized comparability there is a central projection $z \leq z_0$ with $rz \preceq q_1 z \neq 0$. If $rz = 0$, then $z = \sum q_i z$, and so $\mathcal{R}z$ is homogeneous. If $rz \neq 0$, then because q_1 is abelian, it not difficult to see that there is a central projection $z' \leq z$ (the smallest central projection containing rz will do) such that $rz' \sim q_1 z' \neq 0$. In this case $\mathcal{R}z'$ is homogeneous.

To prove the second statement of the theorem, let $\{p_i\}$ be an orthogonal family of pairwise equivalent abelian projections with sum the identity of $\mathcal{R}_n$, and let $u_i \in \mathcal{R}_n$ implement the equivalence $p_1 \sim p_i$. Then

$$x \to \{u_i x u_j^*\}, \qquad x \in \mathcal{R}_n$$

displays the elements of $\mathcal{R}_n$ as "$n \times n$ matrices" over the abelian algebra $\mathcal{A}_n = p_1 \mathcal{R}_n p_1$; in precise terms this means that $\mathcal{R}_n$ is spatially isomorphic to $\mathcal{A}_n \otimes \mathcal{B}(\mathcal{H}_n)$. For details the reader may consult [Sak71, 2.3].

4.6.3

At the opposite extreme from the abelian algebras are the factors. A von Neumann algebra is a **factor** if its center is as small as possible, namely all complex multiples of the identity operator. It is equivalent to say that the algebra cannot be decomposed as the direct sum of von Neumann algebras. Thus, for example, if $\mathcal{R}$ is a type I factor on a Hilbert space $\mathcal{H}$, 4.6.2 implies that there is an isomorphism $\mathcal{H} \cong \mathcal{H}_1 \otimes \mathcal{H}_2$ taking $\mathcal{R}$ to $\mathbf{C}I \otimes \mathcal{B}(\mathcal{H}_2)$; in particular $\mathcal{R}$ is $*$-isomorphic to $\mathcal{B}(\mathcal{H}_2)$. This accounts for the name "factor". We will take up the study of this important notion in a later section. For now we note only that many questions about von Neumann algebras can be reduced to the case of factors. The idea is to use a suitable variant of the concept of a continuous field of C^*-algebras (3.5), namely a **measurable field of von Neumann algebras**. We refer the reader to [KaR86] or [Dix81] for the precise details of the definition. Given a measure space (X, μ) and a measurable field $x \to \mathcal{R}_x, x \in X$, of von Neumann algebras on a fixed separable Hilbert space $\mathcal{H}$, there is an associated von Neumann algebra on $L^2(X, \mu; \mathcal{H})$, denoted by $\int_X \mathcal{R}_x d\mu(x)$ and called the **direct integral** of the field, consisting of elements F of $L^\infty(X, \mu; \mathcal{B}(\mathcal{H}))$ (see 4.3.5) such that $F(x) \in \mathcal{R}_x$ a.e. The result, proved by von Neumann, is that any von Neumann algebra on a separable Hilbert space is a direct integral of factors. In the simplest case, that of a homogeneous algebra $L^\infty(X, \mu) \otimes \mathcal{B}(\mathcal{H})$, the factors $\mathcal{R}_x$ are all equal to $\mathcal{B}(\mathcal{H})$.

4.7 THE TYPE CLASSIFICATION

A projection p in a von Neumann algebra $\mathcal{R}$ is **finite** if, for all $q \in P(\mathcal{R})$, $p \sim q \leq p$ implies that $q = p$. It is not difficult to see that any projection equivalent to a finite one is finite, that abelian projections are finite, and that subprojections and finite orthogonal sums of finite projections are finite (see [Ka51c] for a proof of the last). A von Neumann algebra is **finite** if its identity 1 is a finite projection; it is **semifinite** if 1 can be decomposed as a sum of finite projections (equivalently, every nonzero projection has a nonzero finite subprojection). Thus any type I algebra is semifinite.

A von Neumann algebra is said to be of **type II** if it is semifinite but contains no nonzero abelian projection; it is of **type III** if it contains no nonzero finite projection at all. It is by no means clear that algebras of types II and III exist; some examples are constructed in 4.9 below.

4.7.1

Any von Neumann algebra is uniquely a direct sum

$$\mathcal{R} = \mathcal{R}_{\mathrm{I}} \oplus \mathcal{R}_{\mathrm{II}} \oplus \mathcal{R}_{\mathrm{III}}$$

of algebras of types I, II, and III. For example, to obtain $\mathcal{R}_{\mathrm{I}}$ take a maximal orthogonal family $\{e_i\}$ of abelian projections, and let $z_{\mathrm{I}} = \sum e_i$. Then $z_{\mathrm{I}} \in \mathcal{Z}(\mathcal{R})$; otherwise, $z_{\mathrm{I}}\mathcal{R}z_{\mathrm{I}}^{\perp} \neq \{0\}$, and it would be possible to find, using 4.5.2, a nonzero abelian subprojection of $z^{\perp}$. Hence $\mathcal{R}_{\mathrm{I}} = z_{\mathrm{I}}\mathcal{R}$. Likewise, to split $\mathcal{R}_{\mathrm{II}}$ off $z_{\mathrm{I}}^{\perp}\mathcal{R}$, let z_{II} be the sum of a maximal orthogonal family of finite subprojections of $z_{\mathrm{I}}^{\perp}$.

We have already discussed, in 4.6.2, the decomposition of a type I algebra into its n-homogeneous summands. By grouping together the summands with $n < \infty$, we get the unique maximal finite summand. Any type II algebra also has a unique largest finite summand, obtained by adding up any maximal orthogonal family of finite central projections.

4.7.2

By "Cantor-Bernstein" (property i, 4.5.3) the relation $\preceq$ induces a partial order on the set $D(\mathcal{R}) = P(\mathcal{R})/\sim$ of Murray–von Neumann equivalence classes; by generalized comparability (property ii, 4.5.3) this is a total order if $\mathcal{R}$ is a factor. Recall also that $D(\mathcal{R})$ has the structure of a partial semigroup, obtained by adding orthogonal representatives when these exist. If $\mathcal{R}$ has no (nonzero) finite summand, this is always possible and $D(\mathcal{R})$ is a semigroup. In this case, however, it can never have cancellation. For example, if $\mathcal{H}$ is separable and infinite-dimensional, then

$$D(\mathcal{B}(\mathcal{H})) \cong \{0, 1, 2, \ldots, \infty\}$$

as ordered semigroups, where $\infty + n = \infty$ for all n.

Murray and von Neumann were able to show that for any factor $\mathcal{R}$ on a separable Hilbert space, $D(\mathcal{R})$ is isomorphic (as a partially ordered partial semigroup) to a subsemigroup of $[0, \infty]$ with addition. In fact they showed that

$$D(\mathcal{R}) \cong \begin{cases} \{0, 1, 2, \dots, n\} & \text{if } \mathcal{R} \text{ is type I and finite (type I}_n) \\ \{0, 1, 2, \dots, \infty\} & \text{if } \mathcal{R} \text{ is type I and infinite (type I}_\infty) \\ [0, 1] & \text{if } \mathcal{R} \text{ is type II and finite (type II}_1) \\ [0, \infty] & \text{if } \mathcal{R} \text{ is type II and infinite (type II}_\infty) \\ \{0, \infty\} & \text{if } \mathcal{R} \text{ is type III} \end{cases}$$

In the type II case this result means that one has "continuous dimensions." Thus, for example, any projection can be halved: $p = q + r$ with $q \sim r$. For type III it means that any nonzero projection is equivalent to the identity. The type I factors were described in 4.6.3.

A type II factor is **type II_1** if it is finite, and **type II_∞** if it is infinite.

We will consider the type II case in some detail. A key step in proving the result of Murray and von Neumann in this case is the splitting property just mentioned.

4.7.3 Lemma

Let p be a projection in a type II factor $\mathcal{R}$. Then there are projections q, r in $\mathcal{R}$ with $p = q + r$ and $q \sim r$.

Proof

Once again this is a maximality argument, this time using families of pairs (q_i, r_i) of nonzero equivalent subprojections of p such that the set of all q_i and r_i is pairwise orthogonal. This reduces the assertion to showing that any nonzero projection p contains such a pair. Now p cannot be minimal; otherwise, $p\mathcal{R}p = \mathbf{C}p$ so that p would be abelian. Let q be a proper nonzero subprojection of p. Since we are in a factor, generalized comparability (property ii, 4.5.3) gives $q \preceq p - q$ or $p - q \preceq q$, yielding in either case the required pair. (In fact this last step can be refined a little to prove the lemma for type II nonfactors as well; see [Ka51c, 4.12].)

4.7.4 Theorem

If $\mathcal{R}$ is a finite type II factor on a separable Hilbert space, then there is a unique function $d : P(\mathcal{R}) \to [0, 1]$, called the **dimension function** *of $\mathcal{R}$, such that*

i. $d(1) = 1$.
ii. $p \sim q \Rightarrow d(p) = d(q)$.
iii. $p \perp q \Rightarrow d(p + q) = d(p) + d(q)$.

Moreover d has the additional properties:

iv. $p \preceq q \Leftrightarrow d(p) \leq d(q)$.
v. *d is faithful:* $d(p) = 0 \Rightarrow p = 0$.
vi. *d is* **normal**: $d(\sup p_i) = \sup d(p_i)$ *for any increasing net* $\{p_i\}$.

Outline of the Proof

The idea of the proof is to construct a dyadic scale of projections in terms of which any projection will have an essentially unique binary expansion. By the lemma, $1 = e_1 + e_1'$ with $e_1 \sim e_1'$. Note that by comparison and finiteness, e_1 is unique up to equivalence. Next split $e_1' = e_2 + e_2'$ in the same way. Continuing, we obtain an orthogonal sequence $\{e_i\}$ with $(\sum_{i=1}^{n} e_i)^\perp \sim e_n$ for all n. It is a simple consequence of finiteness that no nonzero projection f has the property that $f \preceq e_n$ for infinitely many n (because there would then be an infinite family of orthogonal copies of f and their sum would be infinite). It follows that $\sum_{i=1}^{\infty} e_i = 1$, since $f = (\sum_{i=1}^{\infty} e_i)^\perp$ satisfies

$$f \leq \left(\sum_{i=1}^{n} e_i \right)^\perp \sim e_n$$

for all n. Now for $p \in P(\mathcal{R})$ let $d(p)$ be the binary decimal $0.\varepsilon_1\varepsilon_2 \ldots$ defined by $\varepsilon_1 = 1$ if $e_1 \preceq p$ and $\varepsilon_1 = 0$ otherwise, $\varepsilon_2 = 1$ if $\varepsilon_1 e_1 + e_2 \preceq p$ and $\varepsilon_2 = 0$ otherwise, and so forth. Then (arguing as for $\sum e_i = 1$) we have $p \sim \sum_{i=1}^{\infty} \varepsilon_i e_i$.

It is clear that d is uniquely determined by conditions i–iii, first on the e_n, then on finite sums of them and on projections equivalent to these, and finally on everything by condition iv (which follows from conditions ii and iii).

4.7.5

The relationship between the type of a tensor product $\mathcal{R} \otimes \mathcal{S}$ and the types of $\mathcal{R}$ and $\mathcal{S}$ is easily determined. For the tensor product to be type I both component algebras must be type I. For it to be type II, one component needs to be type II and the other semifinite (i.e., no type III summand). For example, if $\mathcal{R}$ is type II, so is $\mathcal{R} \otimes \mathcal{B}(\mathcal{H})$. The tensor product of a type III algebra with anything is type III. Finally, a tensor product of finite (semifinite) algebras is finite (semifinite). For the details, see [Tak79, ch.5].

From 4.3.10 we have that the center of $\mathcal{R} \otimes \mathcal{S}$ is the tensor product of the centers; in particular, $\mathcal{R} \otimes \mathcal{S}$ is a factor if and only if both $\mathcal{R}$ and $\mathcal{S}$ are factors.

We note also that for any II_∞ factor $\mathcal{R}$, there is a II_1 factor $\mathcal{R}_1$ such that $\mathcal{R} \cong \mathcal{R}_1 \otimes \mathcal{B}(\mathcal{H})$. To see this, one wants to construct a decomposition $1 = \sum_{i=1}^{\infty} p_i$ into orthogonal equivalent finite projections, for then an argument as in 4.6.2 will give $\mathcal{R} \cong p_1 \mathcal{R} p_1 \otimes \mathcal{B}(\mathcal{H})$. Fix a finite nonzero projection q and a maximal orthogonal family $\{q_i\}$ of projections equivalent to q. By maximality and 4.5.3(ii), the remainder $r = (\sum q_i)^\perp$ satisfies $r \preceq q$. Since the family is infinite, we have $\sum q_i \sim \sum_{i \neq 1} q_i$, and so

$$1 = \sum q_i + r \sim \sum_{n \neq 1} q_i + r \preceq \sum q_i \leq 1$$

Therefore $1 \sim \sum q_i$ by 4.5.3(i), and the existence of the required decomposition follows.

4.7.6

Most of the structure theory to this point has been extended by Kaplansky to the class of Banach $*$-algebras called **AW*-algebras** [Ka51c], and even to suitable $*$-rings. See [Ber72] for a more recent account and additional references. The analysis of $P(\mathcal{R})$ with its equivalence relation $\sim$ has been axiomatized by Loomis [Loo55] and others, and in [Fil65] the quotient $D(\mathcal{R})$ is shown to have the structure of a (generalized) **cardinal algebra**, a concept introduced and studied by Tarski.

4.8 TRACES

In their second paper Murray and von Neumann studied traces on von Neumann algebras [MvN37]. Subsequently Dixmier developed the theory of traces and gave an independent account of of the type classification based on them [Dix81]. In this section we will describe some of these results.

4.8.1

A **trace** on a von Neumann algebra $\mathcal{R}$ is a map $\varphi : \mathcal{R}_+ \to [0, \infty]$ that is additive, homogeneous for scalars $\lambda \geq 0$, and unitarily invariant ($\varphi(u^* a u) = \varphi(a)$ for $a \in \mathcal{R}_+$ and unitary $u \in \mathcal{R}$). A trace is **finite** if it does not take the value ∞; in this case φ extends to a positive linear functional on $\mathcal{R}$ with the "trace property" $\varphi(ab) = \varphi(ba)$ for all $a, b \in \mathcal{R}$. In particular, $\varphi(p) = \varphi(q)$ if p and q are equivalent projections. In fact this is true for any trace [Dix81, 1.6.1].

A trace is **semifinite** if the elements of finite trace are weak operator dense in $\mathcal{R}_+$. A trace φ is **faithful** if $\varphi(a) \neq 0$ for all $a \in \mathcal{R}_+$, $a \neq 0$.

The traces of greatest interest are the **normal** traces—those that satisfy $\varphi(\sup a_i) = \sup \varphi(a_i)$ for any increasing net $\{a_i\}$ in $\mathcal{R}_+$ with a supremum. For a finite trace normality is equivalent to σ-weak continuity.

4.8.2

Murray and von Neumann showed that for any finite type II factor $\mathcal{R}$ the dimension function of 4.7.4 extends to provide a faithful finite trace τ on $\mathcal{R}$ such that $\tau(1) = 1$; they also proved that such a trace is unique and normal. Another description of the trace on a finite type II factor was given by Dixmier: for every $x \in \mathcal{R}$, the norm-closed convex hull of the unitary orbit u^*xu of x meets the center $\mathcal{Z}(\mathcal{R}) = \mathbf{C}1$ in the single point $\tau(x)1$. See also [Yea71].

It follows (using 4.7.5) that any infinite type II factor has a nonzero semifinite trace, unique up to a scale factor and automatically faithful and normal.

The case of the type I infinite factor presents an interesting contrast. It is easily seen that the only semifinite normal trace on $\mathcal{B}(\mathcal{H})$ is the usual one. However, there do exist semifinite nonnormal traces on $\mathcal{B}(\mathcal{H})$, usually referred to as **Dixmier traces** [Dix66]. The construction of these traces begins with two ingredients:

i. A sequence $(a_n) \subset \mathbf{R}$ that is increasing and unbounded and such that $a_n - a_{n+1}$ is decreasing and a_{2n}/a_n has limit 1.
ii. A functional $f \in (\ell^\infty)^*$ that is nonnegative on ℓ^∞_+, vanishes on c_0, and has the "scale invariance" property $f(\lambda^{(2)}) = f(\lambda)$ for all $\lambda \in \ell^\infty$, where $\lambda^{(2)}$ is obtained from λ by repeating each coordinate twice.

For any $K \in \mathcal{K}(\mathcal{H})_+$ let $(\lambda_n(K))$ be the sequence of eigenvalues of K, arranged in decreasing order and counting multiplicities (so ending in a sequence of 0's if K has finite rank), and let $(s_n(K))$ be the sequence of partial sums of the $\lambda_n(K)$. For $T \in \mathcal{B}(\mathcal{H})_+$ the Dixmier trace $\tau(T)$ is then defined by

$$\tau(T) = \begin{cases} f(s_1(T), s_2(T), \ldots) & \text{if } T \in \mathcal{K}(\mathcal{H})_+ \text{ and } (s_n(T)) \in \ell^\infty \\ +\infty & \text{otherwise} \end{cases}$$

The only nonobvious point is that τ is additive, and this follows from the inequalities

$$s_n(A+B) \leq s_n(A) + s_n(B) \leq s_{2n}(A+B)$$

valid for all $A, B \in \mathcal{K}(\mathcal{H})_+$. Note that these traces vanish on the finite rank operators. The Dixmier trace obtained by taking $a_n = \log n$ is used by Connes to construct the quantized calculus needed for noncommutative differential geometry [Con94].

4.8.3

In the general (nonfactor) case, Dixmier obtained the following results:

1. A von Neumann algebra is semifinite if and only if it has a faithful semifinite normal trace.
2. A von Neumann algebra with separable predual is finite if and only if it has a faithful finite normal trace.
3. Any two semifinite normal traces on a factor are proportional.

The uniqueness result 3 does not extend to the nonfactor case. However, if $\mathcal{R}$ is finite there does exist a unique map $T : \mathcal{R} \to \mathcal{Z}(\mathcal{R})$, called the **center-valued trace**, with the following properties:

i. T is linear.
ii. $T(1) = 1$.
iii. $T(ab) = T(ba)$ for all $a, b \in \mathcal{R}$.
iv. $T(a) \geq 0$ for all $a \geq 0$, with equality only for $a = 0$.
v. T is normal.
vi. $T(za) = zT(a)$ for all $z \in \mathcal{Z}(\mathcal{R}), a \in \mathcal{R}$.

If we identify $\mathcal{Z}(\mathcal{R})$ with an L^∞ space, say $\mathcal{Z}(\mathcal{R}) = L^\infty(X, \mu)$, then the normal traces φ on $\mathcal{R}$ have the form

$$\varphi(a) = \int_X T(a) f \, d\mu \qquad \text{for all } a \in \mathcal{R}$$

for some measurable $f \geq 0$ [Dix81, ch. 4].

And, finally, restricting T to the projections in $\mathcal{R}$ gives the **center-valued dimension function** $d : P(\mathcal{R}) \to \mathcal{Z}(\mathcal{R})_+$ on $\mathcal{R}$; it induces an isomorphism of $P(\mathcal{R})/\sim$ and $\{f \in L^\infty(X, \mu) : 0 \le f \le 1\}$ as ordered partial semigroups.

These results can be extended to the semifinite case, except that T is defined only on $\mathcal{R}_+$, and it takes its values in the space of measureable functions $X \to [0, \infty]$.

4.8.4

Type III von Neumann algebras have, by definition, no useful traces. The theory of weights allows them to be studied on an equal footing with the semifinite algebras and has in fact been essential to recent developments that have revolutionized the subject. A **weight** on a von Neumann algebra $\mathcal{R}$ (or on a C^*-algebra) is a map $\mathcal{R}_+ \to [0, \infty]$ that is additive and positively homogeneous. Thus a trace is a weight that is invariant under the action of the unitary group of the algebra. The concept of a weight is very general, and frequently one considers only **semifinite** or **normal** ones (these properties are defined in the same way as for traces).

A simple example of a weight on $\mathcal{B}(\mathcal{H})$ is given by

$$\varphi(T) = \sum \langle T\xi_i, \xi_i \rangle$$

for any set $\{\xi_i\}$ of vectors; in fact the normal weights on any von Neumann algebra are all of this form [Haa75]. For an orthonormal basis this formula gives the usual trace.

In the C^*-algebra context we will meet in Chapter 5 a situation in which it is necessary to restrict to **lower semicontinuous** weights. For a weight φ on a C^*-algebra A, this means that the sets $\{a \in A_+ : \varphi(a) \le t\}$ are norm-closed for all $t \in \mathbf{R}_+$. If A is separable, it is equivalent to say that

$$\varphi(a) = \sup\{\psi(a) : \psi \in A^* \text{ and } 0 \le \psi \le \varphi\}$$

for all $a \in A_+$. Here the inequality $0 \le \psi \le \varphi$ is to be understood in the pointwise sense, as functions on A_+. For a commutative C^*-algebra $C_0(X)$ the functionals $\psi \in A^*$ satisfying $0 \le \psi$ are (via integration) precisely the finite positive Borel measures on X and the lower semi-continuous weights are the unbounded positive Borel measures on X.

Here is an example of a weight that is not lower semicontinuous. Take $X = [0, \infty)$, and let μ be a finite positive Borel measure on the Stone-Čech compactification βX with $\mu(X) = 0$ (e. g., a point mass). Define φ on $C_0(X)_+$

by

$$\varphi(f) = \begin{cases} \int \widetilde{Mf}\, d\mu & \text{if } Mf \text{ is bounded} \\ +\infty & \text{if } Mf \text{ is unbounded} \end{cases}$$

where $f \in C_0(X)_+$, $(Mf)(t) = tf(t)$, and $\widetilde{Mf}$ is the continuous extension of Mf from X to βX. It is easy to see that if $\psi \in C_0(X)^*$ with $0 \leq \psi \leq \varphi$, then $\psi = 0$.

One reason for the importance of weights is their connection with representation theory. This will be explained in the next chapter. For more information, see [Com68] and the references cited there.

4.9 *FACTORS OF TYPES II AND III*

One of the outstanding achievements of Murray and von Neumann was to establish the existence of algebras of types II [MvN36] and III [Neu40]. This is based on their **group measure space construction**, which is in fact a crossed product of the type introduced in 3.8. Let $(G, \alpha, \mathcal{A})$ be a **discrete W^*-dynamical system** with $\mathcal{A}$ abelian (the "measure space"). This means that G is a discrete group, $\mathcal{A}$ is a von Neumann algebra (on a Hilbert space $\mathcal{H}$), and $\alpha : G \to \operatorname{Aut} \mathcal{A}$ is a homomorphism. We need to take G infinite, and this requires that the ideas of 3.8 be modified slightly. The formula for $\rho : \mathcal{A}G \to \mathcal{B}(\ell^2(G, \mathcal{H}))$ given there continues to make sense on the $*$-subalgebra of $\mathcal{A}G$ consisting of the finitely nonzero functions $x : G \to \mathcal{A}$, and the **W^*-crossed product** $\mathcal{A} \rtimes_\alpha G$ is defined to be the von Neumann algebra generated by the operators $\rho(x)$ for such x.

When $\mathcal{A} = \mathbf{C}$ on $\mathcal{H} = \mathbf{C}$ (and α is trivial) the resulting crossed product algebra is the **group von Neumann algebra** $\mathcal{R}(G)$. It acts on $\ell^2(G)$ and is generated by the unitary operators $\lambda_g = \rho(\delta_g)$ described in 3.8.3 (with $\mathcal{H} = \mathbf{C}$).

4.9.1 Theorem

The group von Neumann algebra $\mathcal{R}(G)$ of a discrete group G is finite. It is a factor if and only if the conjugacy classes $\{hgh^{-1} : h \in G\}$ are infinite for all $g \neq e$; in this case it is of type II if $G \neq \{e\}$.

Outline of Proof

The equation $\tau(T) = \langle T\delta_e, \delta_e \rangle$ defines an σ-weakly continuous functional τ on $\mathcal{R}(G)$ that is nonnegative on $\mathcal{R}(G)_+$.

One easily computes that

$$\tau(\rho(x)\lambda_g) = x(g^{-1}) = \tau(\lambda_g(\rho(x))$$

for all $g \in G$ and finitely nonzero $x : G \to \mathbf{C}$, and it follows (since the elements $\lambda_g = \rho(\delta_g)$ generate $\mathcal{R}(G)$) that τ is a trace. It is also true—but more difficult to see—that τ is faithful. For this it is convenient to anticipate a result (5.2.1) from the next chapter, which states that because $\tau|\mathcal{R}(G)_+ \geq 0$, one has

$$|\tau(S^*T)|^2 \leq \tau(S^*S)\tau(T^*T) \qquad \text{for all } S, T \in \mathcal{R}(G)$$

If now $\tau(T) = 0$ for some $T \in \mathcal{R}(G)_+$, then it follows that

$$\tau(T\lambda_g) = \tau(T^{1/2}(T^{1/2}\lambda_g)) = 0$$

for all g, so that (using $\lambda_g\delta_e = \delta_g$ and the fact that τ is a trace)

$$\langle T\delta_g, \delta_h \rangle = \langle \lambda_h^* T\lambda_g\delta_e, \delta_e \rangle = \tau(\lambda_h^* T\lambda_g) = \tau(T\lambda_{h^{-1}g}) = 0$$

for all g, h, and thus $T = 0$.

The second statement of the theorem is a consequence of the following facts about the function $T(g) = \tau(T\lambda_g)$ associated to a given $T \in \mathcal{R}(G)$:

i. $\sum_g |T(g)|^2 < \infty$.
ii. $T(hgh^{-1}) = T(g)$ for all $T \in \mathcal{Z}(\mathcal{R}(G))$ and all $g, h \in G$.

Noting that $\tau(T\lambda_g) = \langle T\delta_e, \delta_{g^{-1}} \rangle$, the sum in the first of these is $\|T\delta_e\|^2$ by Parseval's identity. For the second we have $T\lambda_h = \lambda_h T$, and hence

$$\tau(T\lambda_{hgh^{-1}}) = \tau(T\lambda_h\lambda_g\lambda_{h^{-1}}) = \tau(\lambda_{h^{-1}}T\lambda_h\lambda_g) = \tau(T\lambda_g)$$

Therefore $\tau(T\lambda_g) = 0$ for $g \neq e$, and T is a scalar multiple of the identity.

So in this case $\mathcal{R}(G)$ is a finite factor. By 4.8 it is either type I or type II. But by 4.6.3 and 4.7.2 the finite type I factors are the algebras $\mathcal{B}(\mathcal{H})$ with $\dim \mathcal{H} < \infty$ and are therefore finite-dimensional. Since $\mathcal{R}(G)$ is here infinite-dimensional, it must be of type II.

4.9.2

Examples of groups satisfying the conjugacy class requirement of 4.9.1 include the group S_∞ of all finite permutations of a countably infinite set and the free group F_n on $n \geq 2$ generators. Murray and von Neumann showed that $\mathcal{R}(S_\infty)$ and $\mathcal{R}(F_2)$ are nonisomorphic [MvN36], and subsequently it has been shown that there are uncountably many nonisomorphic finite type II factors [McD69], but little progress has been recorded on the general classification problem for finite type II factors. However, a good deal is known about the subfactors of the hyperfinite factor (defined in 4.10 below).

A long-standing open problem, raised by R. V. Kadison, is whether the free group factors $\mathcal{R}(F_n)$ are all isomorphic. Ideas from Voiculescu's **noncommutative probability theory** have been used to learn a good deal about them. Thus it is known that one has a dichotomy: either these factors are all isomorphic to one another, or else they are pairwise nonisomorphic. Also, for example, $M_2(\mathcal{R}(F_5))$ is isomorphic to $\mathcal{R}(F_2)$. More generally, $M_n(\mathcal{R}(F_{kn^2+1}))$ is isomorphic to $\mathcal{R}(F_{k+1})$; such formulas lead to the construction of free group factors for nonintegral numbers of generators. For details and references, see [Dyk94].

4.9.3

To get examples of algebras of type III, it is necessary to take W^*-crossed products $\mathcal{A} \rtimes_\alpha G$ in which the abelian algebra $\mathcal{A}$ is nontrivial. In fact let $\mathcal{A} = L^\infty(X, \mu)$ where X is a locally compact Hausdorff space, μ is a positive Borel measure on X, and $\alpha : G \to \operatorname{Aut} \mathcal{A}$ is induced by an action of G (also denoted by α) as homeomorphisms of X. The action of G is said to be **free** if the sets $\{x \in X : \alpha_g(x) = x\}$ are of measure zero for all $g \neq e$; it is **ergodic** if $B = \alpha_g(B)$ a.e. for all $g \in G$ is the case for a Borel set B only for B or $X \setminus B$ of measure zero.

Theorem [Neu40]

If the action of G is free and ergodic, then the W^*-crossed product $\mathcal{A} \rtimes_\alpha G$ is a factor.

A proof can be given [Tak79, ch.7] that extends the key idea of 4.9.1. The trace $\tau : \mathcal{R}(G) \to \mathbf{C}$ used in that proof is here replaced by a map $E : \mathcal{A} \rtimes_\alpha G \to \mathcal{A}$ defined as follows: If $\mathcal{H}$ is identified as a subspace of $\ell^2(G, \mathcal{H})$ by sending $\xi \in \mathcal{H}$ to $P\xi = \delta_e \xi$, then $E(t) = P^*TP$ for all $T \in \mathcal{A} \rtimes_\alpha G$. This map can be used to establish the main point of the proof, that if G acts freely, then the canonical image $\rho(\mathcal{A}\delta_e)$ of $\mathcal{A}$ (see 3.8.3) in $\mathcal{A} \rtimes_\alpha G$

is maximal abelian there. Thus $\rho(\mathcal{A}\delta_e)$ contains the center $\mathcal{Z}$ of $\mathcal{A} \rtimes_\alpha G$. But if $\rho(A\delta_e) \in \mathcal{Z}$ for an element $A \in \mathcal{A}$, then in particular A commutes with all λ_g and a short computation shows that this amounts to the requirement that $\alpha_g(A) = A$ for all g. Ergodicity now implies that $A = \lambda I$ for some $\lambda \in \mathbf{C}$ and so $\mathcal{A} \rtimes_\alpha G$ is a factor.

4.9.4

To determine the type of a factor constructed as above, it is necessary to look more carefully at the relation between the measure μ (recall that $\mathcal{A} = L^\infty(X, \mu)$) and the action α. If μ is invariant under α, if $\varphi : \mathcal{A} \to \mathbf{C}$ is the semifinite trace defined by integration with respect to μ, and if E is as in 4.9.3, then it is not difficult to see that $\tau = \varphi \circ E$ is a faithful semifinite σ-weakly continuous trace on $\mathcal{A} \rtimes_\alpha G$. In fact, since replacing μ by an equivalent (i.e., mutually absolutely continuous) measure does not change the factor, to get semifiniteness it is only necessary to know that there exists an invariant Borel measure equivalent to μ. Von Neumann calls such an action **measurable** (but note that this terminology is no longer in use) and we have:

Theorem [Neu40]

Let G act freely and ergodically on $\mathcal{A}$. Then the factor $\mathcal{A} \rtimes_\alpha G$ is of type III if and only if the action is nonmeasurable.

4.9.5

Here is an example of a nonmeasurable action. Let (X, μ) be the real line with Lebesgue measure, let $r > 1$ be a fixed rational number, and let G be the group consisting of all

$$g = (n, s) \in \mathbf{Z} \times \mathbf{Q}$$

acting as homeomorphisms of X by $\alpha_g(x) = r^n x + s$. Let G_0 be the subgroup defined by $n = 0$, and observe that it acts ergodically and leaves μ invariant. Suppose, if possible, that there exists ν equivalent to μ and invariant for G and let $\frac{d\mu}{d\nu}$ be the Radon-Nikodym derivative. Then $\frac{d\mu}{d\nu}$ is invariant for G_0 (because both μ and ν are), hence constant a.e. (because G_0 is ergodic), so that μ is a constant multiple of ν and is therefore invariant for all of G. This contradiction establishes the nonmeasurability of the action of G.

4.10 HYPERFINITE ALGEBRAS

The important class of AF C^*-algebras was introduced in 3.10.4. Here we consider the analogous class of von Neumann algebras, which is perhaps even more fundamental. A von Neumann algebra is called **hyperfinite** if it is generated by an increasing sequence of finite-dimensional $*$-subalgebras [Con94, V.7]. The terminology is due to Dixmier [Dix81], who restricted it to finite factors. The term "matricial" is also used [KaR86]. This idea, introduced by Murray and von Neumann in their study of factors, is connected with some of the most profound results in the theory of von Neumann algebras. They used the term "approximately finite" to designate hyperfinite II_1 factors.

4.10.1

Murray and von Neumann proved the following strong uniqueness result; for a proof we refer the reader to [KaR86, 12.2.1].

Theorem [MvN43]

Any two hyperfinite type II_1 factors are $*$-isomorphic.

Thus, for example, all of the group von Neumann algebras $\mathcal{R}(G)$ considered in 4.9.1 are isomorphic for G locally finite (i.e., a union of an increasing sequence of finite subgroups). Another illustration of the power of this uniqueness result comes from the the UHF algebras $A = \lim_{\to} M_{n_k}(\mathbf{C})$ described in 3.10.1. If A is embedded unitally in some $\mathcal{B}(\mathcal{H})$, then of course the weak operator closure $\mathcal{R}$ will be hyperfinite. The canonical normalized traces on the $M_{n_k}(\mathbf{C})$ are compatible (since n_k divides n_{k+1} for all k) and so give a trace τ on A. If the embedding is such that τ extends to $\mathcal{R}$, it will follow from 4.8.3 that $\mathcal{R}$ is a finite factor—necessarily of type II if A is infinite-dimensional. We will see in 5.3.5 how such an embedding may be constructed for any UHF algebra. Thus the theorem implies that, in striking contrast with the plethora of pairwise nonisomorphic UHF algebras, the corresponding von Neumann algebras are all isomorphic to one another.

We note again that there are plenty of nonhyperfinite type II_1 factors—at least a continuum—of which Murray and von Neumann constructed the first example (see 4.9.2).

4.10.2

It is also true that there is a unique (up to isomorphism) hyperfinite factor of type II_∞. By uniqueness it can be described as $\mathcal{R} \otimes \mathcal{B}(\mathcal{H})$, where $\mathcal{R}$ is the hyperfinite

II_1 factor. However, this uniqueness result is much more difficult than the corresponding result for finite factors; it was obtained by Connes [Con76]. As one indication of the difficulty note that it is by no means obvious that a compression $p\mathcal{R}p$ of a hyperfinite factor $\mathcal{R}$ by a projection p is hyperfinite.

In 1967 R. T. Powers constructed a continuum of nonisomorphic hyperfinite type III factors (see 5.3.6 below). This was the starting point for a series profound results, due to Tomita, Takesaki, Connes, Krieger, Haagerup, and others, that has led to a complete classification of hyperfinite factors. This will be described in more detail in Chapter 9.

Chapter Five

Representations

5.1 STATES

Let A be a C^*-algebra with positive cone A_+ (defined in 2.3). A linear functional $\rho : A \to \mathbf{C}$ is **positive**, written $\rho \geq 0$, if $\rho \mid A_+ \geq 0$. It is a **state** if also $\|\rho\| = 1$. The **state space** of A is the set $S(A)$ of all states. It is a weak*-closed subset of the unit ball of A^* and is therefore weak*-compact by Alaoglu's theorem; we will see (in 5.3.3) that it is convex.

Note that a multiplicative linear functional φ is necessarily positive, since

$$\varphi(a^*a) = \varphi(a^*)\varphi(a) = \overline{\varphi(a)}\varphi(a)$$

by 2.2.1. In general, nonzero multiplicative linear functionals need not exist (e.g., on $M_n(\mathbf{C})$), and it is necessary to work with positive functionals—or even weights—instead.

5.1.1

Positive linear functionals share some of the most important properties of multiplicative functionals. For example, they are automatically bounded. Choose, if possible, a sequence $\{a_n\} \subset \text{ball } A_+$ with $\rho(a_n) \geq 2^n$ for all n and consider $a = \sum_{n=1}^{\infty} 2^{-n} a_n$. Then for all $N \in \mathbf{N}$,

$$\rho(a) \geq \rho\left(\sum_{n=1}^{N} 2^{-n} a_n\right) \geq N$$

a contradiction.

Positive functionals are also **self-adjoint**, in the sense that $\rho(a^*) = \overline{\rho(a)}$ for all $a \in A$. This follows from the **polarization identity**

$$4b^*a = (a+b)^*(a+b) + i(a+ib)^*(a+ib) - (a-b)^*(a-b) \\ - i(a-ib)^*(a-ib)$$

by setting successively $b = 1$ and $a = 1, b = a$.

5.1.2 *Examples*

Positive functionals on $C_0(X)$, X locally compact and Hausdorff, correspond via integration to positive finite Borel measures on X. To include infinite measures, it would be necessary to consider weights (4.8.4). Positive functionals on $M_n(\mathbf{C})$ correspond to elements H of the positive cone $M_n(\mathbf{C})_+$ via the formula

$$\rho(X) = \mathrm{trace}(HX) \qquad \text{for all } X \in M_n(\mathbf{C})$$

In general, if ρ is positive on a C^*-algebra A and $a \in A$, then $x \to \rho(a^*xa)$ is positive on A. If A is a nondegenerate (see 4.2) algebra of operators on a Hilbert space $\mathcal{H}$ and $\xi \in \mathcal{H}$, then the linear functional w_ξ defined by

$$w_\xi(a) = \langle a\xi, \xi\rangle \qquad \text{for all } a \in A$$

is positive. It is a state, called a **vector state**, if $\|\xi\| = 1$.

5.2 *THE CAUCHY–SCHWARZ INEQUALITY*

Let ρ be a positive linear functional or, more generally, a weight on a C^*-algebra A. We then have the following "Cauchy–Schwarz inequality":

5.2.1 *Proposition*

For all $a, b \in A$, $|\rho(a^*b)|^2 \le \rho(a^*a)\rho(b^*b)$.

Proof

This is proved in just the same way as its classical namesake, by expanding

$$\rho((a+\lambda b)^*(a+\lambda b)) \ge 0 \qquad \text{with } \lambda = \frac{t\rho(b^*a)}{|\rho(b^*a)|}, \ t \text{ real}$$

and by using the fact that ρ is self-adjoint (5.1.1).

For the case of a weight, the meaning of the inequality needs some explanation. Notice first that the set

$$A_+^\rho = \{a \in A : \rho(a) < \infty\}$$

is a cone so that ρ extends by linearity in a well-defined manner to its complex linear span A^ρ. The additional assertion of the proposition is that if $a^*a, b^*b \in A_+^\rho$, then $b^*a \in A^\rho$ (the inequality then following as before). But since

$$\begin{aligned}(a + \lambda b)^*(a + \lambda b) &\le (a + \lambda b)^*(a + \lambda b) + (a - \lambda b)^*(a - \lambda b) \\ &= 2(a^*a + |\lambda|^2 b^*b)\end{aligned}$$

for all $a, b \in A$ and $\lambda \in \mathbf{C}$, this is a consequence of the polarization identity (5.1.1) and the fact that $\rho(x) < \infty$ whenever $0 \le x \le y$ and $\rho(y) < \infty$.

5.2.2 *Corollary*

For all $a \in A$,

$$\rho(a^*a) = 0 \iff \rho(ba) = 0 \textit{ for all } b \in A.$$

5.3 *THE GNS REPRESENTATION*

A **representation** of a C^*-algebra A on a Hilbert space $\mathcal{H}$ is a $*$-homomorphism $A \to \mathcal{B}(\mathcal{H})$. Thus the result promised in 2.2.6 and 3.1, the **Gelfand-Naimark theorem**, is the assertion that any C^*-algebra admits a faithful (i.e., injective) representation on some Hilbert space. A key ingredient of this is the **GNS construction** (the acronym is for Gelfand-Naimark-Segal). This important technique is explained in some detail in the next proof.

5.3.1 *Theorem*

To each positive linear functional ρ on a C^*-algebra A corresponds a Hilbert space $\mathcal{H}_\rho$, a vector $\xi_\rho \in \mathcal{H}_\rho$ and a representation $\pi_\rho : A \to \mathcal{B}(\mathcal{H}_\rho)$ such that

i. $\langle \pi_\rho(a)\xi_\rho, \xi_\rho \rangle = \rho(a)$ *for all $a \in A$ and*
ii. $\pi_\rho(A)\xi_\rho$ *is dense in $\mathcal{H}_\rho$.*

Proof

By 5.2.2, $L_\rho = \{a : \rho(a^*a) = 0\}$ is a left ideal of A. By 5.2.1, the formula

$$\langle x + L_\rho, y + L_\rho \rangle = \rho(y^*x)$$

gives a well-defined inner product on the quotient vector space A/L_ρ; let $\mathcal{H}_p$ be the completion. The left regular representation of A on itself, namely $a \to \pi_\rho(a)$ where $\pi_\rho(a) : A \to A$ is left multiplication by a, passes to A/L_ρ because L_ρ is a left ideal. Since (by 2.4.2)

$$\|a(x+L_\rho)\|^2 = \|ax+L_\rho\|^2 = \rho(x^*a^*ax) \leq \|a\|^2\rho(x^*x) = \|a\|^2\|x+L_\rho\|^2$$

for all $a, x \in A$, $\pi_\rho(a)$ extends by continuity to $\mathcal{H}_\rho$ and (as is easily checked) gives a representation $\pi_\rho : A \to \mathcal{B}(\mathcal{H}_\rho)$. If A is unital, take $\xi_\rho = 1 + L_\rho$; conditions i and ii are then immediate. In general, we let $\xi_\rho = \lim(u_\lambda + L_\rho)$, where $\{u_\lambda\}$ is an approximate unit for A (one always exists by 2.5.2). We have $\|u_\lambda - u_\mu + L_\rho\|^2 = \rho((u_\lambda - u_\mu)^2) \leq \rho(u_\lambda - u_\mu)$ for $\lambda \geq \mu$. So to see that the above limit exists, we need to know that $\lim \rho(u_\lambda)$ exists.

5.3.2 Proposition

If $\rho \geq 0$ and $\{u_\lambda\}$ is an approximate unit, then $\rho(u_\lambda) \to \|\rho\|$. In particular, $\|\rho\| = \rho(1)$ if A is unital.

Proof

Since ρ is positive, $\{\rho(u_\lambda)\}$ is an increasing net; since $|\rho(u_\lambda)| \leq \|\rho\|\|u_\lambda\| \leq \|\rho\|$, it is bounded by $\|\rho\|$. Hence $\alpha = \lim \rho(u_\lambda)$ exists, and $\alpha \leq \|\rho\|$. On the other hand, if $\|x\| \leq 1$, then

$$|\rho(u_\lambda x)|^2 \leq \rho(u_\lambda^2)\rho(x^*x) \leq \rho(u_\lambda)\rho(x^*x) \leq \alpha\|\rho\|$$

so that $|\rho(x)|^2 \leq \alpha\|\rho\|$ and $\|\rho\| \leq \alpha$.

5.3.3

The positive functionals on A form a convex subset of A^*, and it follows from 5.3.2 that the state space $S(A)$ is convex as well.

5.3.4

Condition ii of Theorem 5.3.1 states that ξ_ρ is a **cyclic vector** for $\pi_\rho(A)$ or, equivalently, that π_ρ is a **cyclic representation**. Any cyclic representation $\pi : A \to \mathcal{B}(\mathcal{H})$ is **nondegenerate** in the sense that $\pi(A)\xi = 0$ only for $\xi = 0$. This is because if ξ_0 is a cyclic vector, then for all $a \in A$,

$$0 = (\pi(a^*)\xi, \xi_0) = (\xi, \pi(a)\xi_0)$$

so that ξ is orthogonal to the dense subspace $\pi(A)\xi_0$. On the other hand, a routine application of Zorn's lemma shows that any nondegenerate representation is a direct sum of cyclic ones. In this assertion "is" means "is unitarily equivalent to," and the **direct sum of representations** π_λ on Hilbert spaces $\mathcal{H}_\lambda$ is the representation $\pi(a) = \sum \oplus \pi_\lambda(a)$ on $\mathcal{H} = \sum \oplus \mathcal{H}_\lambda$.

We note here that the construction of the GNS representation and the proof of 5.3.1(i) goes through for weights (4.8.4), at least when A is unital. By the argument in 5.2.1, the set

$$A_2^\rho = \{a \in A : \rho(a^*a) < \infty\}$$

is a linear space, and since for all $b \in A$,

$$(ba)^*(ba) = a^*b^*ba \le \|b\|^2 a^*a$$

it is a left ideal. So for the Hilbert space $\mathcal{H}_\rho$ one can take the completion of A_2^ρ / L_ρ, and for $\pi_\rho(b)$ the operator induced by left multiplication by b.

However, to guarantee nondegeneracy, it is necessary to assume some sort of continuity for the weight. In the case of von Neumann algebras, it is usual to assume that the weight is normal, and then it can be proved that the representation is nondegenerate; for C^*-algebras the appropriate hypothesis is that the weight be lower semicontinuous [Ped79, ch.5]. For the example given in 4.8.4, it is easy to check that the GNS representation is the zero representation on a nonzero Hilbert space if the measure is nonzero.

5.3.5 *Examples*

Applied to the positive linear functional $\rho(f) = \int f\,d\mu$ on $A = C_0(X)$, where μ is a finite positive Borel measure on the locally compact Hausdorff space X, the GNS construction yields the multiplication representation of $C_0(X)$ on $L^2(X, \mu)$. If G is a finite group with identity element e and $\mathbf{C}G$ is its complex group algebra, then $\rho(x) = x(e)$ defines a state on $\mathbf{C}G$ for which the GNS construction gives the left regular representation λ described in 3.6.1.

For the next example we return to the construction of the hyperfinite II_1 factor in 4.10.1. Applying the GNS construction to the canonical trace τ on the UHF algebra $A = \lim_{\rightarrow} M_{n_k}(\mathbf{C})$ produces a faithful representation π_τ of A. It follows from conditions i and ii of 5.3.1 that τ extends to the weak operator closure $\mathcal{R}$ of $\pi_\tau(A)$, and hence it is the hyperfinite II_1 factor.

Applying the GNS construction to the trace τ on any II_1 factor M produces what is called the **standard representation** of M. The representation space $\mathcal{H}_\tau$ is denoted $L^2(M, \tau)$, and we note that the range of this representation is already weakly closed.

5.3.6

The idea of fitting traces together as in the previous example can be extended to construct the **Powers factors** [Pow67] mentioned in 4.10.2. Start with the UHF algebra $A = U(2^\infty)$, obtained as the direct limit of the system $M_2(\mathbf{C}) \to M_4(\mathbf{C}) \to M_8(\mathbf{C}) \to \cdots$ with embeddings $a \to \begin{pmatrix} a & 0 \\ 0 & a \end{pmatrix}$, and a number $t \in [0, 1]$. Define states $\phi_n : M_{2^n}(\mathbf{C}) \to \mathbf{C}$ recursively by

$$\phi_1 \begin{pmatrix} a & b \\ c & d \end{pmatrix} = (1 - t)a + td;$$

$$\phi_{n+1} \begin{pmatrix} a & b \\ c & d \end{pmatrix} = (1 - t)\phi_n(a) + t\phi_n(d)$$

These states are compatible with the embeddings and so fit together to define a state ρ_t on A. Let $\pi_t : A \to \mathcal{B}(\mathcal{H}_t)$ be the associated representation and $\mathcal{R}_t$ the von Neumann algebra generated by the range of π_t. Then for $0 < t < \frac{1}{2}$ the $\mathcal{R}_t$ are pairwise nonisomorphic hyperfinite type III factors. (For $t = 0$ we have the I_∞ factor, and for $t = \frac{1}{2}$ the hyperfinite II_1 factor; $\mathcal{R}_t$ and $\mathcal{R}_{1-t}$ are isomorphic for all t.)

5.4 THE GELFAND-NAIMARK THEOREM

We are now ready to state and prove the Gelfand-Naimark theorem.

5.4.1 Theorem [GeN43]

Any C^*-algebra is isometrically $*$-isomorphic to a C^*-subalgebra of some $\mathcal{B}(\mathcal{H})$. If the C^*-algebra is separable, then $\mathcal{H}$ can be taken to be separable.

Proof

A family F of states of a C^*-algebra A is faithful if for any $a \in A_+, a \neq 0$, there exists $\rho \in F$ with $\rho(a) \neq 0$. For such a family, the direct sum representation $\pi = \sum_{\rho \in F} \oplus \pi_\rho$ on $\mathcal{H} = \sum_{\rho \in F} \oplus \mathcal{H}_\rho$ is faithful (because if $\pi_\rho(a) = 0$, then $\rho(a^*a) = \|\pi_\rho(a)\xi_\rho\|^2 = 0$ by 5.3.1, condition i) and hence isometric by 2.2.3. So we need the following lemma:

5.4.2 Lemma

For any $a \in A_+$ there exists $\rho \in S(A)$ such that $\rho(a) = \|a\|$.

Proof

Adjoin a unit, if A is not already unital (1.2). Let $0 \neq a \in A_+$, let $C^*(a)$ be the C^*-subalgebra generated by a and the identity, and define ρ on $C^*(a)$ by evaluation at $\|a\|$ (see 2.2.5). Then $\|\rho\| = \rho(1) = 1$, and we can extend ρ to $\tilde{\rho} \in A^*$, $\|\tilde{\rho}\| = 1$. Somewhat surprisingly, $\tilde{\rho}$ is necessarily positive. This follows from the converse of 5.3.2 (unital case):

$$\rho \in A^* \quad \text{and} \quad \|\rho\| = \rho(1) \Longrightarrow \rho \geq 0$$

Here we can assume $\|\rho\| = 1$; then it will be enough to show that $\rho(x)$ is in the convex hull of $\sigma(x)$ for all $x \in A$ with $x^*x = xx^*$. If the latter statement is not true for some such x, then there must exist $\lambda_0 \in \mathbf{C}$ and $s > 0$ such that $\sigma(x) \subset B(\lambda_0, s)$ but $\rho(x) \notin B(\lambda_0, s)$, where $B(\lambda_0, s)$ is the closed disc centred on λ_0 with radius s. But the former implies $r(x - \lambda_0) \leq s$ and the latter

$$s < |\rho(x) - \lambda_0| = |\rho(x - \lambda_0)| \leq \|x - \lambda_0\|$$

This is a contradiction, for on applying 2.2 and 2.2.2 with $a = x - \lambda_0$ and $A = C^*(x)$ (note that it is commutative), we have

$$r(x - \lambda_0) = \|(x - \lambda_0)\hat{}\|_\infty = \|x - \lambda_0\|$$

Returning to the proof of the theorem, assume that A is separable. Then the representation spaces $\mathcal{H}_\rho$ of 5.3.1 are separable, and there is a sequence $\{a_n\}$ dense in $\{a \in A_+ : \|a\| = 1\}$. For each n choose, as in the first part of the proof, $\rho_n \in S(A)$ with $\rho_n(a_n) = 1$. Then $\sum \oplus \pi_{\rho_n}$ is faithful and $\sum \oplus \mathcal{H}_{\rho_n}$ is separable.

5.4.3 *Corollary*

i. *For any C^*-algebra A, $M_n(A)$ is a C^*-algebra.*
ii. *For any C^*-algebra A, $a \in A_+ \iff \rho(a) \geq 0$ for all $\rho \geq 0$.*

For condition i, if A is faithfully represented on a Hilbert space $\mathcal{H}$, then (as mentioned in 3.2) $M_n(A)$ is naturally represented on the n-fold direct sum of copies of $\mathcal{H}$. Looking at the vector states in a faithful representation of A gives the nontrivial implication in condition ii.

5.4.4

Any positive linear functional ρ on a C^*-subalgebra B of a C^*-algebra A has a positive extension to A with the same norm. To see this, it is enough to extend ρ to the unitization B^+, for then the argument in 5.4.2 serves to extend from B^+ to A^+. Take $\rho^+(b, \lambda) = \rho(b) + \lambda\|\rho\|$. If (u_α) is an approximate unit for B,

$$\rho(bu_\alpha + \lambda u_\alpha) \leq \|\rho\| \|bu_\alpha + \lambda u_\alpha\| \leq \|\rho\| \|(b, \lambda)\|$$

by the definition (1.2) of $\|(b, \lambda)\|$, and consequently

$$|\rho^+(b, \lambda)| = \big|\rho(b) + \lambda\|\rho\|\big| = \lim\big|\rho(bu_\alpha) + \lambda\rho(u_\alpha)\big| \leq \|\rho\| \|(b, \lambda)\|$$

by 5.3.2. Thus $\|\rho^+\| = \|\rho\| = \rho^+(1)$, and therefore $\rho^+ \geq 0$ (see the proof of 5.4.2).

5.4.5

Which representations of a C^*-algebra arise from the GNS construction? Answer: the cyclic ones (up to unitary equivalence). Let $\pi : A \to \mathcal{B}(\mathcal{H})$ be a representation with a unit cyclic vector $\xi \in \mathcal{H}$, and let ρ be the state defined by $\rho(a) = (\pi(a)\xi, \xi)$ for all $a \in A$. If $\pi_\rho : A \to \mathcal{B}(\mathcal{H}_\rho)$ is the corresponding representation, then $\pi(a)\xi \to \pi_\rho(a)\xi_\rho$ defines a unitary $U : \mathcal{H} \to \mathcal{H}_\rho$ such that $\pi(a) = U^*\pi_\rho(a)U$ for all $a \in A$.

5.5 *IRREDUCIBLE REPRESENTATIONS*

A representation $\pi : A \to \mathcal{B}(\mathcal{H})$ of a C^*-algebra is **irreducible** if $\pi(A)$ has no proper closed reducing subspaces. Since $\pi(A)$ is a self-adjoint subalgebra of $\mathcal{B}(\mathcal{H})$, it is equivalent to say that no closed subspace of $\mathcal{H}$, other than $\{0\}$ and $\mathcal{H}$ itself, is a common invariant subspace for all $\pi(a)$, $a \in A$.

5.5.1

The irreducibility of π is also equivalent to the requirement that $T \in \mathcal{B}(\mathcal{H})$ commutes with all $\pi(a)$, $a \in A$, only in the trivial case $T \in \mathbf{C}I$. For T commutes with $\pi(A)$ if and only if T^* does, if and only if Re T and Im T do, if and only if the spectral projections of Re T and Im T do; but a projection P commutes with $\pi(A)$ if and only if its range $P\mathcal{H}$ is invariant for $\pi(A)$.

As a simple consequence we see that the irreducible representations of $C_0(X)$ are all one-dimensional, hence that they are just the multiplicative linear functionals on $C_0(X)$. Another immediate consequence is that the identity representation of $\mathcal{K}(\mathcal{H})$ is irreducible. We will see in 5.6.2 that up to unitary equivalence, this is its only irreducible representation. In particular, this is the case for $M_n(\mathbf{C})$. More generally, any C^*-subalgebra of $\mathcal{K}(\mathcal{H})$ is isomorphic to a c_0-direct sum $\sum_\lambda \oplus\mathcal{K}(\mathcal{H}_\lambda)$, and so its irreducible representations correspond to the summands [Arv76, §1.4].

5.6 PURE STATES

Any irreducible representation is cyclic (any nonzero vector will do), hence arises from a state. Which states yield irreducible representations?

5.6.1 Theorem

For any state ρ of a C^*-algebra A, π_ρ is irreducible if and only if ρ is an extreme point of the state space $S(A)$.

Lemma

For any positive $\rho, \sigma \in A^*$, $\sigma \le \rho$ if and only if there exists $H \in \mathcal{B}(\mathcal{H}_\rho)$ such that $0 \le H \le I$, H commutes with $\pi_\rho(A)$ and

$$\sigma(a) = \langle H\pi_\rho(a)\xi_\rho, \xi_\rho\rangle \qquad \textit{for all } a \in A$$

Proof

The equation $\varphi(\pi_\rho(a)\xi_\rho, \pi_\rho(b)\xi_\rho) = \sigma(b^*a)$ defines a positive sesquilinear form φ on a dense subspace of $\mathcal{H}_\rho$. It is well-defined and bounded, since

$$\begin{aligned} |\varphi(\pi_\rho(a)\xi_\rho, \pi_\rho(b)\xi_\rho)|^2 &\leq \sigma(b^*b)\sigma(a^*a) \\ &\leq \rho(b^*b)\rho(a^*a) \\ &= \|\pi_\rho(b)\xi_\rho\|^2 \|\pi_\rho(a)\xi_\rho\|^2 \end{aligned}$$

by 5.2.1. Hence there exists $H \in \mathcal{B}(\mathcal{H}_\rho)$ such that

$$\sigma(b^*a) = \langle H\pi_\rho(a)\xi_\rho, \pi_\rho(b)\xi_\rho\rangle \qquad \text{for all } a, b \in A$$

Put $b = u_\lambda$, an approximate unit, and take limits. To see that H commutes with $\pi_\rho(A)$, observe that for all a, b, $c \in A$ we have

$$\begin{aligned} \langle H\pi_\rho(c)\pi_\rho(a)\xi_\rho, \pi_\rho(b)\xi_\rho\rangle &= \sigma(b^*(ca)) = \sigma((c^*b)^*a) \\ &= \langle \pi_\rho(c)H\pi_\rho(a)\xi_\rho, \pi_\rho(b)\xi_\rho\rangle \end{aligned}$$

Proof of the Theorem

If $\rho = s\sigma + (1 - s)\tau$ for some $\rho, \sigma, \tau \in S(A)$ and $0 \leq s \leq 1$, then, by the lemma, $s\sigma(a) = (H\pi_\rho(a)\xi_\rho, \xi_\rho)$ for some H commuting with $\pi_\rho(A)$. Thus, if π_ρ is irreducible, we have $H = \lambda I$, and so $s\sigma = \lambda\rho$. According as this is zero or not, we conclude that $\rho = \tau$ or $\rho = \sigma$.

Conversely, let ρ be an extreme point of $S(A)$, and consider $H \in \mathcal{B}(\mathcal{H}_\rho)_+$ that commutes with $\pi_\rho(A)$. Then the equations

$$\varphi(a) = \langle H\pi_\rho(a)\xi_\rho, \xi_\rho\rangle, \qquad \psi(a) = \langle (I - H)\pi_\rho(a)\xi_\rho, \xi_\rho\rangle$$

define positive linear functionals on A and

$$\rho = \varphi + \psi = \|\varphi\|(\|\varphi\|^{-1}\varphi) + \|\psi\|(\|\psi\|^{-1}\psi)$$

Of course $\|\varphi\|^{-1}\varphi$ and $\|\psi\|^{-1}\psi$ are states, but also

$$\|\varphi\| + \|\psi\| = \lim \varphi(u_\lambda) + \lim \psi(u_\lambda) = \lim \rho(u_\lambda) = 1$$

Hence $\rho = \|\varphi\|^{-1}\varphi$, and so $H = \|\varphi\| I$. It follows from 5.5.1 that π_ρ is irreducible.

5.6.2

A **pure state** of A is an extreme point of $S(A)$. Notation: $P(A)$. Thus the theorem asserts that the conditions π_ρ irreducible, ρ pure, and $\rho \in P(A)$ are equivalent. Since $S(A)$ is convex (5.3.3) and weak*-compact (5.1), by the Krein-Milman theorem it is the weak*-closed convex hull of $P(A)$. From this and 5.4.2 it follows that $\rho(a^*a) = 0$ for all $\rho \in P(A)$ occurs only for $a = 0$, so that $\sum \oplus\{\pi_\rho : \rho \in P(A)\}$ is a faithful representation. Even $P(A)$ is larger than necessary: We need only one representative of each unitary equivalence class of irreducible representations.

5.6.3

The pure states of $\mathcal{K}(\mathcal{H})$ are exactly the vector states w_ξ, $\xi \in \mathcal{H}$. One way to see this is to use the fact that functionals $\varphi \in \mathcal{K}(\mathcal{H})^*$ correspond to trace class operators T via the formula $\varphi(K) = \text{trace}(TK)$, $K \in \mathcal{K}(\mathcal{H})$ (see [Mur90, 4.2.1]). In this correspondence φ is a state if and only if $T \geq 0$ with trace 1. Moreover, using the spectral decomposition of the compact positive operator T, it can be seen that φ is pure if and only if T is a projection of rank 1. If $T\xi = \xi$ with $\|\xi\| = 1$, then $\varphi = w_\xi$ as asserted.

It follows from this that if π is an irreducible representation of $\mathcal{K}(\mathcal{H})$, then π is unitarily equivalent to the identity representation. By 5.6.1, π is unitarily equivalent to π_ρ for some pure state ρ. But $\rho = w_\xi$ for some $\xi \in \mathcal{H}$, and a check of the GNS construction shows that $K\xi \to K + L_\rho (K \in \mathcal{K}(\mathcal{H}))$ defines a unitary operator $U : \mathcal{H} \to \mathcal{H}_\rho$ such that $U^*\pi_\rho(K)U = K$ for all $K \in \mathcal{K}(\mathcal{H})$.

5.6.4

The **Kadison transitivity theorem** [KaR83, 5.4.3] asserts that if a C^*-subalgebra $\mathcal{A}$ of $\mathcal{B}(\mathcal{H})$ acts irreducibly on $\mathcal{H}$, then for any vectors $\xi_1, \ldots, \xi_n, \eta_1, \ldots, \eta_n \in \mathcal{H}$ with the ξ_i linearly independent, there exists $T \in \mathcal{A}$ such that $T\xi_i = \eta_i$ for all i. This has an interesting implication for the GNS representation $\pi_\rho : A \to \mathcal{B}(\mathcal{H}_\rho)$ associated with a pure state ρ of the C^*-algebra A. In the notation of 5.3.1, the dense linear subspace A/L_ρ of $\mathcal{H}_\rho$ is invariant for the irreducibly acting algebra $\mathcal{A} = \pi_\rho(A)$, and therefore, by transitivity, $A/L_\rho = \mathcal{H}_\rho$ and is consequently already complete.

5.6.5

The existence of irreducible representations can be used to show that any finite-dimensional C^*-algebra is $*$-isomorphic to a direct sum of matrix algebras.

Arguing inductively, it suffices to show that A has some $M_n(\mathbf{C})$ as a direct summand. But A, being finite-dimensional, has a minimal nonzero ideal, say I, and it will be enough to show $I \cong M_n(\mathbf{C})$, since a unital ideal of a ring is necessarily a direct summand. Let $\pi : I \to \mathcal{B}(\mathcal{H})$ be irreducible. Then π is injective (since I is minimal) and $\mathcal{H}$ is finite-dimensional (since π is cyclic and I is finite-dimensional); hence

$$I \cong \pi(I) \supset \mathcal{K}(\mathcal{H}) = \mathcal{B}(\mathcal{H})$$

by 3.11.2, and so $I \cong \mathcal{B}(\mathcal{H})$.

5.7 THE UNIVERSAL REPRESENTATION

Let A be a C^*-algebra with state space $S(A)$. The **universal representation** of A is the direct sum $\pi_S = \sum \oplus\{\pi_\rho : \rho \in S(A)\}$ acting on $\mathcal{H}_S = \sum_\rho \oplus \mathcal{H}_\rho$. It is faithful (by 5.4.2), and the weak operator closure $\pi_S(A)''$ of its range is called the **enveloping von Neumann algebra** of A [Ped79, 3.8]. The algebra $\pi_S(A)$ is frequently identified with A, and then A is a subalgebra of the enveloping algebra A''. Moreover any state of A is then a vector state.

5.7.1

The universal representation has the property that any nondegenerate representation $\pi : A \to \mathcal{B}(\mathcal{H}_\pi)$ extends uniquely to an σ-weakly continuous surjection $A'' \to \pi(A)''$. This is clear if $\pi = \pi_\rho$ for some $\rho \in S(A)$; the desired extension is simply $T \to T|\mathcal{H}_\rho$ (note that $\mathcal{H}_\rho$ reduces $\pi_S(A)$ and therefore also $\pi_S(A'')$). If π is cyclic with a unit cyclic vector ξ, the assertion follows from the fact that π and π_ρ are unitarily equivalent, where ρ is the vector state associated with ξ. The result for general π now follows by decomposing it into cyclic pieces.

5.7.2

For any C^*-algebra A, the canonical Banach space embedding $A \hat{\to} A^{**}$ of A is its second dual extends to an isomorphism $A'' \to A^{**}$ of Banach spaces. This map may conveniently be described as the adjoint of a suitable map $e : A^* \to (A'')_{*}$, where $(A'')_*$ is the space of σ-weakly continuous linear functionals on A'' so that (see 4.4) the dual space of $(A'')_*$ is A''. In fact e takes any continuous functional on A to its (unique) σ-weakly continuous extension to all of A''. This extension certainly exists for any state of ρ of A:

As mentioned above, ρ is a vector state and so is defined automatically on all of $\mathcal{B}(\mathcal{H}_S)$. The general case then follows from the fact that any element of A^* is a linear combination of (four) states [Ped79, 3.1 and 3.2]. Evidently e is an isometry. It is surjective since $e(\varphi|A) = \varphi$ for any $\varphi \in (A'')_*$. Finally, if $a \in A$, then for all $\psi \in A^*$,

$$e^*(a)\psi = ae(\psi) = \psi(a) = \hat{a}(\psi)$$

and so $e^*(a) = \hat{a}$.

5.8 EXTENSION OF REPRESENTATIONS

Next we consider the possibility of extending a representation given on an ideal or a subalgebra to the whole algebra.

5.8.1 Proposition

Any representation $\pi : I \to \mathcal{B}(\mathcal{H})$ of an ideal I of a C^*-algebra A extends to a representation of A. If π is nondegenerate, then the extension is unique. If π is faithful and I is an essential ideal, then the extension is faithful.

Proof

For $a \in A$ define $\pi(a)$ on $\pi(I)\mathcal{H}$ by

$$\pi(a)\pi(i)\xi = \pi(ai)\xi \qquad \text{for all } i \in I \text{ and } \xi \in \mathcal{H}$$

This is bounded and well-defined: For $\xi \in \mathcal{H}$,

$$\begin{aligned} \|\pi(ai)\xi\|^2 &= \langle \pi(i^*a^*ai)\xi, \xi\rangle \\ &\le \|a\|^2 \langle \pi(i^*i)\xi, \xi\rangle \\ &= \|a\|^2 \|\pi(i)\xi\|^2 \end{aligned}$$

making use of 2.4.2. The definition of $\pi(a)$ is completed by setting $\pi(a) = 0$ on the orthogonal complement of $\pi(I)\mathcal{H}$. If π is nondegenerate, then $\pi(I)\mathcal{H}$ is dense in $\mathcal{H}$, and the extension is unique. If π is faithful and I is essential, then $\pi(a) = 0$ implies that $\pi(ai) = 0$ for all $i \in I$. So $ai = 0$ for all $i \in I$, and hence $a = 0$.

5.8.2

For a C^*-algebra A let $M(A)$ be its multiplier algebra, defined in 3.12.2. We can use 5.8.1 to obtain the characterization promised in 3.12.4.

Corollary

If $\pi : A \to \mathcal{B}(\mathcal{H})$ is faithful and nondegenerate, then π extends uniquely to a faithful representation $\tilde{\pi} : M(A) \to \mathcal{B}(\mathcal{H})$ and

$$\tilde{\pi}(M(A)) = \{T \in \mathcal{B}(\mathcal{H}) : T\pi(A), \pi(A)T \subset \pi(A)\}$$

Proof

By 3.12.3, condition i, the multiplier algebra $M(A)$ contains A as an essential idea, so by 5.8.1 the faithful extension $\tilde{\pi}$ exists and is unique. The identification of $\tilde{\pi}(M(A))$ is almost immediate.

In particular, the multiplier algebra of a nondegenerate C^*-subalgebra $\mathcal{A}$ of $\mathcal{B}(\mathcal{H})$ can be identified as asserted in 3.12.4:

$$M(\mathcal{A}) = \{T \in \mathcal{B}(\mathcal{H}) : T\mathcal{A}, \mathcal{A}T \subset \mathcal{A}\}$$

Note that $M(\mathcal{A}) \subset \mathcal{A}''$; this results from the fact that $\mathcal{A}''$ is the weak operator closure of $\mathcal{A}$, so $I \in \mathcal{A}''$ and if $T\mathcal{A} \subset \mathcal{A}$ then $T\mathcal{A}'' \subset \mathcal{A}''$.

5.8.3

If $\pi : B \to \mathcal{B}(\mathcal{H})$ is a representation of a C^*-subalgebra B of a C^*-algebra A, then it is possible, at the expense of enlarging the Hilbert space, to "extend" π to a representation σ of A. Precisely, there exist Hilbert spaces $\mathcal{H}'' \supset \mathcal{H}'$ and a representation $\sigma : A \to \mathcal{B}(\mathcal{H}'')$ such that $\mathcal{H}'$ is invariant for $\sigma(B)$ and $\sigma|\mathcal{H}'$ on B is unitarily equivalent to π. In proving this, it is enough to consider the situation where π is cyclic, A is unital, and B is a unital subalgebra (adjoin a unit to A, if necessary, and extend π in the obvious way to the subalgebra generated by B and the unit). Let ξ be a cyclic vector for π, and let $\rho \in A^*$ be an extension of norm 1 of the vector state w_ξ so that (see the proof of 5.4.2) ρ is a state. If $(\mathcal{H}_\rho, \xi_\rho, \pi_\rho)$ is the corresponding GNS representation, then $\mathcal{H}'' = \mathcal{H}_\rho$, $\sigma = \pi_\rho$ and $\mathcal{H}' = \overline{\pi_\rho(B)\xi_\rho}$ have the required properties.

5.9 GROUP REPRESENTATIONS

Recall from 3.6.2 that a unitary representation of a topological group G is a homomorphism $g \to U_g$ of G into the group of unitary operators on a Hilbert space $\mathcal{H}$ such that $g \to U_g\xi$ is continuous for all $\xi \in \mathcal{H}$. If G is locally compact, then, as will be shown later, the Haar measure on G can be used to associate to any representation U of G a nondegenerate representation π_U of the group C^*-algebra $C^*(G)$ (introduced in 3.6) such that the von Neumann algebra generated by the range of π_U is the same as that generated by the range of U. This means that to a great extent the representation theory of G is mirrored by that of $C^*(G)$. In particular, it follows that any locally compact group G has an adequate supply of irreducible representations, in the sense that for any $g \in G$ other than the identity element there is an irreducible representation U such that $U_g \neq I$.

5.9.1

Ideally one would like to be able to classify all representations of all locally compact groups. This program can be divided into two parts: Express any representation in terms of irreducible ones, and classify the latter. In the case of a finite group, the results are that any representation decomposes as a direct sum of irreducible ones in an essentially unique manner, and that all irreducibles arise in the decomposition of the left regular representation (3.6.1). Most of this carries over to the case of compact groups, the main difference being that now there can be infinitely many inequivalent irreducibles. For example, for the circle group the regular representation decomposes as the direct sum of the irreducible representations $e^{is} \to e^{ins}$ for $n \in \mathbf{Z}$, every irreducible is unitarily equivalent to exactly one of these, and every representation is a direct sum of irreducibles.

5.9.2

To go beyond the compact case direct sums no longer suffice; a generalization to **direct integrals** is needed [KaR86, ch.14]. The idea is to index the "summands" by the points of a measure space rather than by a discrete set. To get some feeling for this, consider the group $\langle \mathbf{R}, + \rangle$ with the usual topology. Let U be the representation on $L^2(\mathbf{R})$ defined by

$$(U_t\xi)(s) = e^{ist}\xi(s) \qquad \text{for all } \xi \in L^2(\mathbf{R}) \text{ and } s, t \in \mathbf{R}$$

This is unitarily equivalent (via the Fourier transform) to the regular representation of $\mathbf{R}$, but it has the advantage of making the invariant subspaces visible. They are the subspaces $L^2(S)$, where S is a measurable subset of $\mathbf{R}$, consisting of the elements of $L^2(\mathbf{R})$ that vanish outside S. In particular, no nonzero subrepresentation of U is irreducible, and consequently U cannot be a direct sum of irreducibles. In fact the irreducibles are the one-dimensional representations $t \to e^{ist}$ for $t \in \mathbf{R}$, and U is the direct integral (with respect to Lebesgue measure) of them. Likewise an arbitrary representation of $\langle \mathbf{R}, + \rangle$ decomposes as a direct sum of subrepresentations, each of which is a direct integral of the above irreducibles with respect to some Borel measure on $\mathbf{R}$. Since a direct integral counts each irreducible at most once, the direct sum is needed to account for multiplicities. In general, the decomposition of a representation of a group G makes use of (measure classes of) Borel measures on the spectrum of $\widehat{C^*(G)}$, a space that will be described in the next chapter. The introduction of direct integrals means that such decompositions always exist, but the price to be paid is that uniqueness does not generally hold (but see 6.5.3). For more details, see [Dix77, Part II] or [HeR63].

Chapter Six

Structure Theory

6.1 INTRODUCTION

We now return to the study of general C^*-algebras. We want to see to what extent the ideas of the commutative theory (2.2) can be extended. Some of the principal results will be described, mostly without proof.

6.2 THE SPECTRUM OF A C^*-ALGEBRA

The key concept for the study of commutative algebras was that of a multiplicative linear functional. In general, a C^*-algebra A may have no nonzero multiplicative linear functionals. For example, this is the case for $A = \mathcal{K}(\mathcal{H})$ when $\dim \mathcal{H} > 1$. Is there a suitable replacement? The results of 5.5 and 5.6 suggest several possibilities. One has already been identified in 5.6.2: the **pure state space** $P(A)$, equipped with the relative weak*-topology. Another is the **spectrum** of A, defined as the space $\hat{A}$ of unitary equivalence classes $[\pi]$ of nonzero irreducible representations π of A. The GNS construction provides a surjection $P(A) \to \hat{A}$, and $\hat{A}$ is given the quotient topology. Finally, borrowing from general ring theory, one can consider the **structure space** or **primitive ideal space** $\text{Prim}(A)$, consisting of the kernels of the irreducible representations of A. It is provided with the **Jacobson** or **hull-kernel topology**: $S \subset \text{Prim}(A)$ is closed if every primitive ideal that contains the intersection $\cap\{K : K \in S\}$ is already in S. There is an obvious surjection $\hat{A} \to \text{Prim}(A)$; it can be shown [Dix77, 3.4] that the topology of $\hat{A}$ is the weakest making this map continuous. The surjections

$$P(A) \to \hat{A} \to \text{Prim}(A)$$

are evidently bijections when A is commutative.

The pure state space is usually considered to be "too big," in the sense that it tends not to be locally compact. For example, $P(\mathcal{K}(\mathcal{H}))$ can be identified (using 5.6.3) as the S^1-orbit space of the unit sphere $\|\xi\| = 1$ of $\mathcal{H}$ with the relative weak topology. (It can also be noted that $\widehat{\mathcal{K}(\mathcal{H})}$ and $\text{Prim}(\mathcal{K}(\mathcal{H}))$ are singletons.)

On the other hand, the primitive ideal space tends to be "too small." Thus, if A is simple, so that every nonzero irreducible representation has kernel $\{0\}$, $\text{Prim}(A)$ is a singleton. And yet there is a tremendous variety of simple C^*-algebras; to study these, it will be necessary to take account of irreducible representations and not just ideals.

6.2.1 ***Theorem*** [Dix77, 3.3.8]

The spaces $\hat{A}$ *and* $\text{Prim}(A)$ *are locally compact.*

This does not mean that these spaces are Hausdorff. It is clear (from the definition of the Jacobson topology) that $\text{Prim}(A)$ is always T_0; consequently $\hat{A}$ will be T_0 if and only if $\hat{A} \to \text{Prim}(A)$ is a bijection. But $\text{Prim}(A)$ need not be Hausdorff. If

$$A = \{f \in C([0, 1], M_2(\mathbf{C})) : f(0) \text{ is diagonal}\}$$

then (up to unitary equivalence) any irreducible representation is either two dimensional and is given by evaluation at a point of $(0, 1]$, or else is one dimensional and is given by evaluation of the $(1, 1)$ or the $(2, 2)$ entry at 0. Thus $\hat{A}$ and $\text{Prim}(A)$ can be identified as sets with $\{0_1, 0_2\} \cup (0, 1]$, and it is not difficult to see that the topology on $(0, 1]$ is the usual one but that any neighborhood of 0_1 or 0_2 contains an interval of the form $(0, \varepsilon)$.

6.2.2

The topology of $\hat{A}$ can be described in a third way. For any cardinal number n, let $\mathcal{H}_n$ be a fixed Hilbert space of dimension n, and let $\text{Irr}_n(A)$ denote the space of irreducible representations of A on $\mathcal{H}_n$. Let $\text{Irr}_n(A)$ have the pointwise weak operator topology, and denote by $\hat{A}_n$ the image of the obvious map $\text{Irr}_n(A) \to \hat{A}$. Then the quotient and relative topologies on the $\hat{A}_n$ are the same [Dix77, 3.5.8]. For $n < \infty$ it is also true that $\hat{A}_n$ is Hausdorff and that $\bigcup\{\hat{A}_k \mid k \leq n\}$ is closed in $\hat{A}$ [Dix77, 3.6].

6.2.3

Mimicking the Gelfand transform, the idea now is to study A by representing it as functions of some sort on $\hat{A}$. One possibility would be to associate to $x \in A$ the map $[\pi] \to \pi(x)$; this involves choosing representatives and has the additional difficulty that in general the image algebras $A/\ker\pi$ depend on π. The map $\hat{x} : \hat{A} \to \mathbf{R}_+$ defined by $\hat{x}([\pi]) = \|\pi(x)\|$ avoids these problems and is sometimes useful. It is always lower semicontinuous and vanishes at infinity, and is continuous for all $x \in A$ exactly when $\hat{A}$ is Hausdorff [Dix77, 3.3]. The map $[\pi] \to \operatorname{trace}\pi(x)$, for $x \in A_+$, has also been studied. It takes values in $\mathbf{R}_+ \cup \{+\infty\}$ and is always lower semicontinuous [Dix77, 3.5.9].

6.3 HOMOGENEOUS ALGEBRAS

To go much further, it is necessary to restrict the generality. After the commutative algebras the next simplest class would appear to be the ***n*-homogeneous algebras**, for which all irreducible representations have the same finite dimension n. Thus the 1-homogeneous algebras are precisely the commutative ones. These algebras were introduced by Kaplansky [Ka51a], who showed that if A is unital and n-homogeneous then $\hat{A}$ is compact and Hausdorff (so that $\hat{A} \to \operatorname{Prim}(A)$ is bijective). He also proved that A is naturally a module over $C(\hat{A})$ as follows. For every $x \in A$ and $f \in C(\hat{A})$, there exists a unique $fx \in A$ such that

$$(fx)(P) = f(P)x(P) \qquad \text{for all } P \in \operatorname{Prim}(A) = \hat{A}$$

A consequence is that the center of A is isomorphic to $C(\hat{A})$. This was subsequently generalized to any C^*-algebra: $\mathcal{Z}(M(A))$ and $C_b(\operatorname{Prim}(A))$ are isomorphic [DaH68].

The simplest example of an n-homogeneous C^*-algebra A with $\hat{A}$ a given locally compact Hausdorff space X is $A = C_0(X, M_n(\mathbf{C}))$. This can be viewed as the space $\Gamma_0(B)$ of continuous sections, vanishing at infinity, of the trivial M_n-bundle $B = X \times M_n$. In fact every n-homogeneous algebra arises as the algebra of continuous sections of some M_n-bundle [Fel61]. Moreover $\Gamma_0(B) \cong \Gamma_0(B')$ if and only if $B \cong B'$ up to a homeomorphism of X. When A is unital a somewhat more elementary description is possible.

6.3.1 *Theorem*

If A is a unital n-homogeneous C^*-algebra, then there exist an integer $N \geq n$ and a projection-valued function $P \in C(\hat{A}, M_N(\mathbf{C}))$ of constant dimension n such that $A \cong PC(\hat{A}, M_N(\mathbf{C}))P$.

Sketch of the Proof

Assume, as above, that A is the algebra $\Gamma(B)$ of sections of an M_n-bundle B over a compact Hausdorff space $X(= \hat{A})$. The first point is that there is then a Hermitian vector bundle E over X such that $\mathrm{Hom}(E, E)$, the C^*-algebra of bundle morphisms $E \to E$ (see 3.4), is $*$-isomorphic to $\Gamma(B)$. Roughly speaking, E is obtained from B by replacing each of the fibers M_n by $\mathbf{C}^n$, and the isomorphism $\Gamma(B) \to \mathrm{Hom}(E, E)$ by associating to a section $s : X \to B$ the morphism $h_s : E \to E$ given locally by $h_s(x, v) = (x, s(x)v)$. The second point is that because X is compact, there will exist a vector bundle F over X such that $E \oplus F$ is isomorphic to a trivial bundle $X \times \mathbf{C}^N$ [Ati67, 1.4.14]. Now for each $x \in X$, let $P(x)$ denote the orthogonal projection of $\mathbf{C}^N$ onto its subspace V_x, where $\{x\} \times V_x$ is the image of the fiber E_x. Then $P \in C(X, M_N)$ and

$$PC(X, M_N)P \cong \mathrm{Hom}(E, E) \cong \Gamma(B)$$

6.3.2

These results make it possible to describe, in certain cases, all the isomorphism classes of n-homogeneous C^*-algebras A having a given spectrum $\hat{A}$. For example, if $\hat{A}$ is contractible, then it follows that $A \cong C_0(\hat{A}, M_n)$ because there are no nontrivial bundles over a contractible space.

Krauss and Lawson (KrL74] have considered homogeneous algebras A with $\hat{A} = SX$, where X is compact and SX is the **suspension of X**, obtained by collapsing $X \times \{1\}$ and $X \times \{-1\}$ to distinct points in $X \times [-1, 1]$. The reason for considering spectra of this form is that isomorphism classes of M_n-bundles over SX correspond bijectively to homotopy classes of maps $X \to PU_n$, where $PU_n = U_n/U_1$ is the projective unitary group. In particular, isomorphism classes of M_n-bundles over the sphere $S^{k+1} = S(S^k)$ correspond to the elements of the homotopy group $\pi_k(PU_n)$. Thus up to isomorphism the only n-homogeneous C^*-algebra A with $\hat{A} = S^1$ is $C(S^1, M_n)$. Again, since $\pi_1(PU_n)$ is the cyclic group of order n, there are exactly n nonisomorphic n-homogeneous algebras A with $\hat{A} = S^2$. The corresponding M_n-bundles over S^2 may be obtained by using the maps $\alpha_i : S^1 \to \mathrm{Aut}\, M_n$, where for

$i = 0, 1, \ldots, n-1,$

$$\alpha_i(z) = \mathrm{Ad}\begin{pmatrix} z^i & & & 0 \\ & 1 & & \\ & & \ddots & \\ 0 & & & 1 \end{pmatrix}$$

to glue the trivial M_n-bundles over the upper and lower hemispheres of S^2 together along the equator S^1.

Kaplansky [Kap70] gives a geometric description of the real analogues of these when $n = 2$. Define $P : S^2 \to M_3(\mathbf{R})$ by letting $P(s)$ be projection on the subspace of $\mathbf{R}^3$ orthogonal to s. Then $A = PC(S^2, M_3(\mathbf{R}))P$ is 2-homogeneous with $\hat{A} = S^2$. However, A is not isomorphic to $C(S^2, M_2(\mathbf{R}))$: The latter contains nontrivial projections, but A does not because S^2 does not admit a nonvanishing vector field.

6.4 CCR ALGEBRAS

It is natural to consider next those C^*-algebras A for which $\pi(a)$ is a compact operator for all irreducible π and all $a \in A$. Such an algebra is said to be a **CCR algebra** [Ka51a], the name indicating completely continuous (i.e., compact) representations. Of this name Dixmier [Dix64] said "... la terminologie m'a paru peu esthétique"; he proposed **liminaire** (in English, **liminal** or **liminary**) instead, to which Kaplansky [Kap70] replied, "I stand ready to yield to Dixmier's 'liminaire,' except I confess that I do not know what it means." Pedersen added that the modern terms "do not mean anything, which may be more esthetic," and that the use of acronyms should not be tolerated in mathematics [Ped79].

6.4.1

The n-homogeneous algebras ($n < \infty$) are CCR, as is $\mathcal{K}(\mathcal{H})$ (by 5.6.3). In general, the spectrum of a CCR algebra need not be Hausdorff, as the example following 6.2.1 shows. Dupré [Dup79] has analyzed CCR algebras having irreducible representations of exactly two distinct dimensions. Ideals and quotients of CCR algebras are CCR, but the converse is false: $\mathcal{K}(\mathcal{H}) + \mathbf{C}I$ (the unitization of $\mathcal{K}(\mathcal{H})$) is an example.

6.4.2

A class of C^*-algebras between the n-homogeneous and the CCR algebras has attracted a good deal of attention. In any C^*-algebra A the elements $a \in A^+$ for which the map $[\pi] \to \operatorname{tr} \pi(a)$ is finite and continuous on $\hat{A}$ form the positive part of an ideal of A, and A is said to be a **continuous trace algebra** if this ideal is dense in A. Such algebras are necessarily CCR with Hausdorff spectrum [Dix77, 4.5.3]. Moreover one can associate to such an algebra an isomorphism invariant $\delta(A) \in H^3(\hat{A}, \mathbf{Z})$, called the **Dixmier-Douady invariant** [Dix77, 10.7]. With suitable restrictions this is in fact a complete invariant. For example, if X is a second countable locally compact Hausdorff space of finite dimension, then the isomorphism classes of $\aleph_0$-homogeneous continuous trace separable C^*-algebras with spectrum X correspond bijectively, via the Dixmier-Douady invariant, to the elements of $H^3(X, \mathbf{Z})$. With respect to the tensor product over $C_b(X)$, this bijection is in fact an isomorphism of groups [Dix77, 10.9.5].

The proof of this result is highly technical but the basic idea is quite straightforward. Let $\mathcal{H}$ be a Hilbert space of dimension $\aleph_0$, and let $\mathcal{U}(\mathcal{H})$ be its unitary group. An element of $H^3(X, \mathbf{Z})$ gives rise to an open cover $\{U_i\}$ of X and continuous maps

$$\gamma_{ij} : U_i \cap U_j \to \mathcal{U}(\mathcal{H}) \quad \text{and} \quad \lambda_{ijk} : U_i \cap U_j \cap U_k \to \mathbf{C}$$

the latter of modulus 1, such that $\gamma_{ij}\gamma_{jk} = \lambda_{ijk}\gamma_{ik}$. From this data a C^*-algebra A with $\hat{A} = X$ can be constructed by using the γ_{ij} in a more or less evident manner to piece together the trivial local algebras $C(U_i, \mathcal{K}(\mathcal{H}))$. In particular, the zero element of $H^3(X, \mathbf{Z})$ corresponds to $C(X, \mathcal{K}(\mathcal{H}))$.

6.5 TYPE I C^*-ALGEBRAS

A C^*-algebra is **GCR** or **postliminary** if the image of every nonzero representation contains some nonzero compact operator. Any GCR algebra has a (possibly transfinite) composition series with CCR quotients [Dix77, 4.3.4]. It is equivalent to require continuous trace quotients [Dix77, 4.5.5]. Any C^*-algebra A has a largest CCR ideal, namely

$$I_0 = \{a \in A : \pi(a) \text{ is compact for all irreducible } \pi\}$$

Repeating with A/I_0, and so forth, one arrives eventually at the largest GCR ideal of A. This will be A itself if A is GCR; at the opposite extreme, where this ideal is 0, A is **NGCR** or **antiliminary**.

6.5.1

Any CCR algebra is GCR. The converse is false, as the algebra $\mathcal{K}(\mathcal{H}) + \mathbf{C}I$ shows. For an ideal I of an algebra A, it is true in general that A is GCR if and only if I and A/I are. The largest GCR ideal of $\mathcal{B}(\mathcal{H})$ is $\mathcal{K}(\mathcal{H})$, and consequently the Calkin algebra $\mathcal{B}(\mathcal{H})/\mathcal{K}(\mathcal{H})$ is NGCR. A simple algebra is either isomorphic to some $\mathcal{K}(\mathcal{H})$ or else is NGCR. This follows from an important property of GCR algebras: If $\pi : A \to \mathcal{B}(\mathcal{H}_\pi)$ is an irreducible representation of a GCR algebra A, then $\pi(A) \supset \mathcal{K}(\mathcal{H}_\pi)$. To see this, note that by definition $\pi(A) \cap \mathcal{K}(\mathcal{H}_\pi) \neq \{0\}$ and hence that $\pi(A) \supset \mathcal{K}(\mathcal{H}_\pi)$ by 3.11.2. If, in addition, A is simple, then $A \cong \pi(A) = \mathcal{K}(\mathcal{H}_\pi)$. In particular, it follows that any infinite-dimensional UHF algebra (3.10.1) is NGCR.

6.5.2

A representation $\pi : A \to \mathcal{B}(\mathcal{H}_\pi)$ of a C^*-algebra is said to be **type I** if the von Neumann algebra $\pi(A)''$ generated by the range of π is type I. For example, any irreducible representation is type I. Kaplansky showed that every representation of a GCR algebra is type I [Ka51b]. A C^*-algebra for which every representation is type I is called a **type I C^*-algebra**; thus GCR implies type I. The difficult converse was established by Glimm [Gli61] in the separable case and by Sakai [Sak71] generally. Note that this terminology can be misleading: $\mathcal{B}(\mathcal{H})$ is a type I von Neumann algebra but is not a type I C^*-algebra.

6.5.3

A locally compact group G will be called **type I** if its group C^*-algebra $C^*(G)$ is type I (see 3.6 and 5.9; the definition of $C^*(G)$ in the general case is given in 8.3). This is a large and important class of groups. For example, abelian groups are type I, as the range of any representation generates an abelian von Neumann algebra. Any representation of a compact group is a direct sum of finite-dimensional irreducible representations [Dix77, 15.1]; consequently compact groups are type I. Not all discrete groups are type I. Indeed, as was shown in 4.9.1, the left regular representation of a countable discrete group having no finite nontrivial conjugacy classes is type II. Finally, Harish-Chandra has shown that any connected semisimple Lie group is type I [Dix77, 15.5].

6.5.4

Are the irreducible representations of separable C^*-algebras classifiable? Here "classifiable" means "classifiable up to unitary equivalence," so one might have in mind the existence, for an algebra A, of a countable family $\{f_n\}$ of real-valued functions on $\hat{A}$ (the "unitary invariants") that separate the points of $\hat{A}$. Invariants that are constructible in any sense would have some regularity properties, but continuity would be too much to expect as it is possible to imagine even CCR algebras A for which there are no nonconstant continuous functions $f : \hat{A} \to \mathbf{R}$. A more reasonable requirement is that the f_n are **Borel functions**; this means that they are measurable with respect to the σ-algebras on $\hat{A}$ and $\mathbf{R}$ generated by the open sets. When this is possible A is said to be **smooth** or to have a **smooth dual**.

An important consequence of smoothness is the existence of a Borel cross section of the map $\pi \to [\pi]$ from irreducible representations to unitary equivalence classes of them. Then, by means of the direct integral theory introduced in 5.9.2, one can associate to any Borel measure μ on $\hat{A}$ a representation of A by "integrating" this Borel cross section with respect to μ. The resulting representation is independent, up to unitary equivalence, of the particular cross section; and equivalent (i.e., mutually absolutely continuous) measures yield unitarily equivalent representations.

Is the resulting map, from equivalence classes of Borel measures on $\hat{A}$ to unitary equivalence classes of representations of A, one-to-one? Mackey [Mac57] showed that it is for type I (= GCR) algebras and, in the process, was led to conjecture that for separable C^*-algebras, type I and smooth dual are equivalent conditions. This was established by Glimm [Gli61]:

6.5.5 Theorem

Let A be a separable C^*-algebra. Then the following are equivalent:

i. *A is GCR.*
ii. *A is type I.*
iii. *A has smooth dual.*
iv. *A has no representations of type II.*
v. *A has no representations of type III.*

For more details, the reader should consult [Arv76] and [Ped79].

Chapter Seven

Tensor Products

7.1 NUCLEARITY

In 3.7 it was shown that for $*$-algebras A and B the algebraic tensor product $A \otimes B$ is in a natural way a $*$-algebra. Moreover it was indicated that if A and B are C^*-algebras, then there is at least one C^*-norm on $A \otimes B$, namely the spatial norm. Here we recall that a C^*-norm is a norm on a $*$-algebra that satisfies the C^*-identity. Note that completeness is not assumed. A C^*-algebra A is said to be **nuclear** if for every C^*-algebra B there is only one C^*-norm on $A \otimes B$. The class of nuclear algebras has turned out to be very important. It has good closure properties, contains most commonly occurring algebras, and admits useful alternative characterizations. For a survey of the subject to 1980, see [Lan82]; [KaR86] and [Mur90] provide detailed accounts.

7.1.1

In general, a norm on the algebraic tensor product of normed algebras may bear no relationship to the norms on the factors. A reasonable compatibility requirement is that

$$\|a \otimes b\| = \|a\| \|b\| \qquad \text{for all } a, b$$

and such a norm is called a **cross-norm**. In the case of C^*-algebras it is true, as we will see later, that C^*-norms are necessarily cross-norms. This results from the following interesting facts: The algebraic tensor product of C^*-algebras always admits a largest C^*-norm $\|\cdot\|_{\max}$ and a smallest C^*-norm $\|\cdot\|_{\min}$, and both are cross-norms. It is a remarkable and far from obvious fact that $\|\cdot\|_{\min}$ is precisely the spatial norm described in 3.7.4. There is also a useful description of $\|\cdot\|_{\max}$; these topics are developed below.

7.1.2

The definition of nuclearity is not particularly convenient in practice. For example, it is not immediately clear that there are any nuclear algebras at all. Even the nuclearity of $\mathbf{C}$ is of some interest, as it amounts to the assertion that for every C^*-algebra B there is only one C^*-norm on B. To see this, let $\|\cdot\|$ and $\|\cdot\|_\alpha$ be C^*-norms on a $*$-algebra B, assume that B is complete in the first of these, and let B_α be the completion of B in the second. Then $\mathrm{id} : B \to B_\alpha$ is a morphism of C^*-algebras and so has closed range, whence $B = B_\alpha$ by 2.5.5, and consequently the norms are the same by 2.1.4. This argument applies to any C^*-algebra A for which $A \otimes B$ is always a C^*-algebra. This is the case for finite dimensional A; for example, $M_n(\mathbf{C}) \otimes B$ is $*$-isomorphic to $M_n(B)$ by 3.7.5, and the latter is a C^*-algebra by 5.4.2. Thus finite-dimensional algebras are nuclear.

7.1.3

We will see later that commutative algebras are nuclear, and in fact any type I C^*-algebra is nuclear. Nuclearity is preserved by direct limits; thus UHF and AF algebras are nuclear (but recall that infinite dimensional UHF algebras are never type I (6.5.1)). These results are due to Takesaki [Tak64]. On the other hand, $\mathcal{B}(\mathcal{H})$ is not nuclear when $\mathcal{H}$ is infinite-dimensional. Indeed the nuclear von Neumann algebras are precisely the finite direct sums of n-homogeneous algebras $M_n(\mathcal{Z})$, n finite [Was76].

7.2 THE SPATIAL TENSOR PRODUCT

In this section we will construct the spatial tensor product and, by considering states on it, establish its main properties.

7.2.1 Proposition

For Hilbert spaces $\mathcal{H}$ and $\mathcal{K}$, there is a unique bilinear map

$$\phi : \mathcal{B}(\mathcal{H}) \times \mathcal{B}(\mathcal{K}) \to \mathcal{B}(\mathcal{H} \otimes \mathcal{K})$$

such that $\phi(S, T)(\xi \otimes \eta) = S\xi \otimes T\eta$ for all $S \in \mathcal{B}(\mathcal{H})$, $T \in \mathcal{B}(\mathcal{K})$, $\xi \in \mathcal{H}$, and $\eta \in \mathcal{K}$. Moreover the induced linear map

$$\Phi : \mathcal{B}(\mathcal{H}) \otimes \mathcal{B}(\mathcal{K}) \to \mathcal{B}(\mathcal{H} \otimes \mathcal{K})$$

on the algebraic tensor product is an injective $*$-homomorphism.

Proof

For $S \in \mathcal{B}(\mathcal{H})$ and $T \in \mathcal{B}(\mathcal{K})$, the composition

$$\mathcal{H} \times \mathcal{K} \underset{S\times T}{\longrightarrow} \mathcal{H} \times \mathcal{K} \to \mathcal{H} \otimes \mathcal{K}$$

of the canonical map following $S \times T$ is bilinear and hence induces a linear map

$$\phi(S, T) : \mathcal{H} \otimes \mathcal{K} \to \mathcal{H} \otimes \mathcal{K}$$

such that $\phi(S, T)(\xi \otimes \eta) = S\xi \otimes T\eta$ for all $\xi \in \mathcal{H}$ and $\eta \in \mathcal{K}$. Evidently

$$\phi(S, T) = \phi(S, I)\phi(I, T)$$

so to prove that $\phi(S, T)$ is bounded, it will be enough to consider the factors. Let $u = \sum \xi_i \otimes \eta_i$ be an element of $\mathcal{H} \otimes \mathcal{K}$. By reasoning as in 3.7.2 with the Gram-Schmidt orthogonalization process, the η_i can be assumed to be pairwise orthogonal. Then the $S\xi_i \otimes \eta_i$ are pairwise orthogonal, and

$$\begin{aligned} \|\phi(S, I)u\|^2 &= \sum \|S\xi_i \otimes \eta_i\|^2 = \sum \|S\xi_i\|^2\|\eta_i\|^2 \\ &\leq \|S\|^2 \sum \|\xi_i\|^2\|\eta_i\|^2 = \|S\|^2\|u\|^2 \end{aligned}$$

The other factor is treated similarly, and it follows that $\|\phi(S, T)\| \leq \|S\|\|T\|$. So $\phi(S, T)$ extends to a bounded linear operator on the completion $\mathcal{H} \otimes \mathcal{K}$, which we also denote by $\phi(S, T)$. Since ϕ is bilinear, it induces a linear map

$$\Phi : \mathcal{B}(\mathcal{H}) \otimes \mathcal{B}(\mathcal{K}) \to \mathcal{B}(\mathcal{H} \otimes \mathcal{K})$$

that takes $S \otimes T$ to $\phi(S, T)$. It is routine to check that the relations

$$\phi(S, T)\phi(S', T') = \phi(SS', TT')$$

$$\phi(S, T)^* = \phi(S^*, T^*)$$

are valid, and it follows that Φ is a $*$-homomorphism. To see that Φ is injective, define, for each $\zeta \in \mathcal{K}$, an operator $R_\zeta \in \mathcal{B}(\mathcal{H} \otimes \mathcal{K}, \mathcal{H})$ by the requirement

$$R_\zeta(\xi \otimes \eta) = \langle \eta, \zeta \rangle \xi \qquad \text{for all } \xi \in \mathcal{H} \text{ and } \eta \in \mathcal{K}$$

This operator has the property that

$$R_\zeta \Phi\left(\sum S_i \otimes T_i\right)(\xi \otimes \eta) = \sum \langle T_i \eta, \zeta \rangle S_i \xi$$

If $\Phi(\sum S_i \otimes T_i) = 0$, then assuming (as we can by 3.7.2) that the S_i are linearly independent, we conclude that the T_i all vanish, hence that $\sum S_i \otimes T_i = 0$.

7.2.2 *Remarks*

It is customary to regard Φ as an inclusion and to write $S \otimes T$ for the operator $\phi(S, T)$ as well as for the element of $\mathcal{B}(\mathcal{H}) \otimes \mathcal{B}(\mathcal{K})$.

The inequality $\|S \otimes T\| \leq \|S\|\|T\|$ obtained in the course of the proof is in fact an equality: If $\xi \in \mathcal{H}$ and $\eta \in \mathcal{K}$ are unit vectors, then so is $\xi \otimes \eta$ and

$$\|S\xi\|\|T\eta\| = \|S\xi \otimes T\eta\| = \|(S \otimes T)(\xi \otimes \eta)\| \leq \|S \otimes T\|$$

The von Neumann algebra tensor product of $\mathcal{B}(\mathcal{H})$ and $\mathcal{B}(\mathcal{K})$ was defined in 4.3.10 as the weak operator closure of their algebraic tensor product, viewed as operators on $\mathcal{H} \otimes \mathcal{K}$. This von Neumann algebra tensor product is in fact all of $\mathcal{B}(\mathcal{H} \otimes \mathcal{K})$. This may be seen by noting that the algebraic tensor product is generated by $\mathcal{B}(\mathcal{H}) \otimes \mathbf{C}I_{\mathcal{K}}$ and $\mathbf{C}I_{\mathcal{H}} \otimes \mathcal{B}(\mathcal{K})$, so that its commutant is

$$(\mathcal{B}(\mathcal{H}) \otimes \mathbf{C}I_{\mathcal{K}})' \cap (\mathbf{C}I_{\mathcal{H}} \otimes \mathcal{B}(\mathcal{K}))' = \mathbf{C}I_{\mathcal{H}} \otimes \mathbf{C}I_{\mathcal{K}} = \mathbf{C}I_{\mathcal{H}\otimes\mathcal{K}}$$

7.2.3 *Lemma*

For any morphisms $\pi : A \to C$ and $\sigma : B \to D$ of C^*-algebras, there is a unique $*$-homomorphism of algebraic tensor products

$$\pi \otimes \sigma : A \otimes B \to C \otimes D$$

such that $(\pi \otimes \sigma)(a \otimes b) = \pi(a) \otimes \sigma(b)$ for all $a \in A$ and $b \in B$. Moreover, if π and σ are injective, so is $\pi \otimes \sigma$.

Proof

The existence and uniqueness of $\pi \otimes \sigma$ as a linear map follows from the bilinearity of the composition

$$A \times B \xrightarrow[\pi \times \sigma]{} C \times D \to C \otimes D$$

That $\pi \otimes \sigma$ is multiplicative and $*$-preserving is a consequence of the corresponding properties of π and σ. The injectivity assertion is obtained by writing $u \in \ker(\pi \otimes \sigma)$ in the form $\sum a_i \otimes b_i$ with the a_i linearly independent.

7.2.4

The category of R-modules, R a commutative ring, has a tensor product with many of the properties of the tensor product of vector spaces. However, the injectivity part of 7.2.3 is not true in this generality, even for $R = \mathbf{Z}$. For example, if $\pi : \mathbf{Z} \to \mathbf{Z}$ is multiplication by 2 and $\sigma : \mathbf{Z}_2 \to \mathbf{Z}_2$ is the identity, then $\pi \otimes \sigma = 0$. An R-module F is called **flat** if $\pi \otimes \mathrm{id}_F$ is injective for every injection π of R-modules. It is true generally that the tensor product of surjectives is surjective, and consequently if F is flat and

$$0 \to U \to V \to W \to 0$$

is an exact sequence of R-modules, then

$$0 \to U \otimes F \to V \otimes F \to W \otimes F \to 0$$

is exact too.

7.2.5

As a consequence of 7.2.1 and 7.2.3, we see that $A \otimes B$ admits a C^*-norm for any C^*-algebras A and B. If $\pi : A \to \mathcal{B}(\mathcal{H}_\pi)$ and $\sigma : B \to \mathcal{B}(\mathcal{H}_\sigma)$ are faithful representations, then there is an injective $*$-homomorphism $\pi \otimes \sigma$ from $A \otimes B$ into $\mathcal{B}(\mathcal{H}_\pi \otimes \mathcal{K}_\sigma)$ such that

$$(\pi \otimes \sigma)(a \otimes b) = \pi(a) \otimes \sigma(b) \qquad \text{for all } a \in A, b \in B$$

This fact may be used to transfer the operator norm to $A \otimes B$:

$$\|u\|_* = \|(\pi \otimes \sigma)(u)\| \qquad \text{for all } u \in A \otimes B$$

It is called the **spatial norm** on $A \otimes B$. As noted in 7.2.2, this is a cross-norm. The next goal is to show that it is independent of the choice of π and σ.

7.2.6 Lemma

If A is a C^*-subalgebra of $\mathcal{B}(\mathcal{H})$, then $S(A)$ is the weak-closed convex hull of the vector states.*

Proof

Let Ω be the set of vector states ω (5.1.2). It has the property that if $a \in A$ is self-adjoint and $\omega(a) \geq 0$ for all $\omega \in \Omega$, then $a \geq 0$. The polar of Ω, namely the set

$$\Omega^0 = \{a \in A_{sa} : \omega(a) \leq 1 \text{ for all } w \in \Omega\}$$

where A_{sa} denotes the self-adjoint part of A, therefore consists of all $a \in A_{sa}$ with $a \leq 1$. Consequently the bipolar Ω^{00} is the set of all $g \in (A_{sa})^*$ such that $g(a) \leq 1$ for all $a \leq 1$. The bipolar theorem asserts that Ω^{00} is the weak*-closed convex hull of Ω and 0. Now, if $f \in S(A)$, then $f \in \Omega^{00}$, and so it is a weak*-limit of a net $\{g_\alpha\}$, where each g_α is of the form

$$g = \sum t_i g_i, \quad g_i \in \Omega, \quad t_i \geq 0, \quad \sum t_i \leq 1$$

Then $g_\alpha(1) \to 1$. Thus, if one takes $f_\alpha = g_\alpha / g_\alpha(1)$, then f_α is a convex combination of vector states and $f_\alpha \to f$ weak*.

7.2.7

For $f \in A^*$ and $g \in B^*$ (where A and B are C^*-algebras), we recall (3.7.1) that $f \otimes g$ is the linear map $A \otimes B \to \mathbf{C}$ defined by

$$(f \otimes g)(a \otimes b) = f(a)g(b) \qquad \text{for all } a \in A, b \in B$$

If f and g are states, it is not difficult to see that $f \otimes g$ is a state on $A \otimes B$, in the sense that it satisfies $(f \otimes g)(u^*u) \geq 0$ for any $u \in A \otimes B$ and has norm 1 with respect to any C^*-norm on $A \otimes B$ for which it is continuous. For positivity, note that if $u = \sum a_i \otimes b_i$, then

$$(f \otimes g)(u^*u) = \sum f(a_i^*a_j)g(b_i^*b_j) = \operatorname{tr}(FG) \geq 0$$

since the matrices $F = (f(a_i^*a_j))$ and $G = (g(b_i^*b_j))$ are positive definite. As for continuity, it is a key fact—one unfortunately not easy to prove—that $f \otimes g$ is continuous in *any* C^*-norm $\|\cdot\|_\alpha$ on $A \otimes B$. Thus there is always an inclusion $S(A) \times S(B) \subset S(A \otimes_\alpha B)$, where $A \otimes_\alpha B$ denotes the completion. This need not be an equality.

7.2.8 Lemma.

Given C^*-algebras A and B, a C^*-seminorm $\|\cdot\|_\alpha$ on $A \otimes B$, and states $f \in S(A)$ and $g \in S(B)$, the following statements are equivalent:

i. $|(f \otimes g)(u)| \leq \|u\|_\alpha$ *for all* $u \in A \otimes B$.
ii. $\|(\pi \otimes \sigma)(u)\| \leq \|u\|_\alpha$ *for all* $u \in A \otimes B$, *where* π *and* σ *are the GNS representations corresponding to* f *and* g.

Proof

If ξ and η are the cyclic vectors for π and σ and $u = \sum a_i \otimes b_i$, then a short computation shows that

$$(f \otimes g)(u) = \langle(\pi \otimes \sigma)(u)(\xi \otimes \eta), \xi \otimes \eta\rangle \qquad (*)$$

Therefore property ii implies property i. On the other hand, vectors of the form

$$(\pi \otimes \sigma)(v)(\xi \otimes \eta), \qquad v \in A \otimes B$$

are dense in $\mathcal{H}_\pi \otimes \mathcal{H}_\sigma$ and by $(*)$

$$\langle(\pi \otimes \sigma)(u)(\pi \otimes \sigma)(v)(\xi \otimes \eta), (\pi \otimes \sigma)(v)(\xi \otimes \eta)\rangle = (f \otimes g)(v^*uv)$$

Therefore proving that property i implies property ii amounts to proving that property i implies

$$|(f \otimes g)(v^*uv)| \leq \|u\|_\alpha \|(\pi \otimes \sigma)(v)(\xi \otimes \eta)\|^2$$

The hypothesis allows $f \otimes g$ to be extended to a positive linear functional on the completion $A \otimes_\alpha B$. Then

$$h(x) = (f \otimes g)(v^*xv), \qquad x \in A \otimes_\alpha B$$

defines a positive linear functional h, so

$$\|h\| = h(1 \otimes 1) = (f \otimes g)(v^*v)$$

and

$$|(f \otimes g)(v^*uv)| = h(u) \leq (f \otimes g)(v^*v)\|u\|_\alpha$$

(If A or B is nonunital, use approximate units and 5.3.2.) But

$$(f \otimes g)(v^*v) = \|(\pi \otimes \sigma)(v)(\xi \otimes \eta)\|^2$$

by $(*)$ and the proof is complete.

7.2.9 Theorem

Let $\pi_0 : A \to \mathcal{B}(\mathcal{H}_0)$ and $\sigma_0 : B \to \mathcal{B}(\mathcal{K}_0)$ be faithful representations of C^*-algebras A and B. Then for all $u \in A \otimes B$,

$$\|(\pi_0 \otimes \sigma_0)(u)\| = \sup\{\|(\pi \otimes \sigma)(u)\| : \pi, \sigma \text{ cyclic}\}$$

In particular this quantity, the spatial norm, is independent of the choice of faithful representations.

Proof

One inequality is easy: If $\pi_0 = \sum \oplus \pi_i$ and $\sigma_0 = \sum \oplus \sigma_j$ are decompositions into cyclic representations, then

$$\pi_0 \otimes \sigma_0 = \sum \oplus(\pi_i \otimes \sigma_j) \quad \text{and} \quad \|(\pi_0 \otimes \sigma_0)(u)\| = \sup \|(\pi_i \otimes \sigma_j)(u)\|$$

For the other inequality it will simplify the notation a little to assume, as we can with no loss of generality, that A and B are C^*-subalgebras of $\mathcal{B}(\mathcal{H})$ and $\mathcal{B}(\mathcal{K})$ and that π_0 and σ_0 are the identity representations. Fix a unit vector $\eta \in \mathcal{K}$, and consider the set

$$X = \{f \in S(A) : |f \otimes \omega_\eta(u)| \leq \|u\| \qquad \text{for all } u \in A \otimes B\}$$

where ω_η is the vector state. It is weak*-closed, convex, and contains the vector states, so $X = S(A)$ by 7.2.6. Now fix $f \in S(A)$, and repeat this reasoning with

$$Y = \{g \in S(B) : |(f \otimes g)(u)| \leq \|u\| \text{ for all } u \in A \otimes B\}$$

to conclude that $|f(\otimes g)(u)| \leq \|u\|$ for all $f \in S(A)$, $g \in S(B)$, and $u \in A \otimes B$. Therefore $\|(\pi \otimes \sigma)(u)\| \leq \|u\|$ for all cyclic representations π and σ by 7.2.8 and 5.4.5.

7.2.10

It is also true that the spatial norm on $A \otimes B$ can be expressed in terms of the irreducible representations of A and B:

$$\|u\|_* = \sup\{\|(\pi \otimes \sigma)(u)\| : \pi, \sigma \text{ irreducible}\}, \qquad u \in A \otimes B$$

One inequality follows from 7.2.9, since irreducible representations are cyclic. For the other, the reasoning in 7.2.9 applies with vector states replaced by pure states and 7.2.6 by 5.6.2.

7.2.11

The proof that the spatial norm is the smallest C^*-norm on $A \otimes B$ is lengthy and will only be sketched. The details can be found in [Tak79]. The main ingredients of the proof are as follows:

1. The topology of $\hat{A}$ is both (a) the quotient topology of $P(A) \to \hat{A}$ and (b) the weak topology of $\hat{A} \to \text{Prim}\, A$ (see 6.2).
2. If $(f, g) \in P(A) \times P(B)$, then $f \otimes g \in P(A \otimes_* B)$, where the (temporary) notation $\otimes_*$ denotes the completion in the spatial norm.
3. Commutative C^*-algebras are nuclear.
4. Any pure state on a C^*-subalgebra extends to a pure state on the whole algebra.
5. If $h \in P(A \otimes_\alpha B)$ and $f = h|_{A\otimes 1} \in P(A)$, then $g = h|_{1\otimes B} \in P(B)$ and $h = f \otimes g$.

These facts are combined to show that the set

$$Q = \{(\pi, \sigma) \in \hat{A} \times \hat{B} : \|(\pi \otimes \sigma)(u)\| \le \|u\|_\alpha \text{ for all } u \in A \otimes B\}$$

is all of $\hat{A} \times \hat{B}$ (so that $\|u\|_* \le \|u\|_\alpha$ for all $u \in A \otimes B$ by 7.2.10). If not, since Q is closed by 1(a) and 7.2.8, then by 1(b) there exist nonzero $a_0 \in A_+$, $b_0 \in B_+$ such that $(\pi \otimes \sigma)(a_0 \otimes b_0) = 0$ for all $(\pi, \sigma) \in Q$. Choose $f \in P(A)$ and $g \in P(C^*(b_0))$ with $f(a_0)g(b_0) \neq 0$. By 2, $f \otimes g$ is a pure state of $A \otimes_* C^*(b_0)$. This can be identified as a C^*-subalgebra of $A \otimes_\alpha B$ by 3, so 4 allows $f \otimes g$ to be extended to a pure state of h of $A \otimes_\alpha B$. Then $h = f \otimes g'$ by 5, where g' is an extension of g. If (π, σ') and (ξ, η) are the GNS representations and cyclic vectors associated to (f, g'), then $(\pi, \sigma') \in Q$

by 7.2.8, and yet

$$\langle (\pi \otimes \sigma')(a_0 \otimes b_0)(\xi \otimes \eta), \xi \otimes \eta \rangle = f(a_0)g(b_0) \neq 0$$

a contradiction.

Notation

The spatial norm will be written $\|\cdot\|_{\min}$. This suggests its minimality and is the generally preferred notation. The completion in the spatial norm will be written $A \otimes_{\min} B$.

7.2.12

An important property of the spatial norm is that if $\pi : A \to C$ and $\sigma : B \to D$ are $*$-homomorphisms of C^*-algebras, then the $*$-homomorphism

$$\pi \otimes \sigma : A \otimes B \to C \otimes D$$

of 7.2.3 extends by continuity to a $*$-homomorphism

$$\pi \otimes \sigma : A \otimes_{\min} B \to C \otimes_{\min} D$$

of the completions. This follows from 7.2.9 by composing π and σ with faithful representations of C and D. The same reasoning shows that the extended $\pi \otimes \sigma$ is injective if both π and σ are. In particular, if $A \subset C$ and $B \subset D$, then $A \otimes_{\min} B \subset C \otimes_{\min} D$ isometrically.

7.2.13

Let $0 \to A \xrightarrow{i} B \xrightarrow{p} C \to 0$ be a short exact sequence of C^*-algebras, let D be a C^*-algebra, and consider the exactness of the tensored sequence

$$0 \longrightarrow A \otimes_{\min} D \xrightarrow{i \otimes \mathrm{id}} B \otimes_{\min} D \xrightarrow{p \otimes \mathrm{id}} C \otimes_{\min} D \longrightarrow 0$$

From the above these maps exist and $i \otimes \mathrm{id}$ is injective. It is clear that $p \otimes \mathrm{id}$ is surjective and that $I = \mathrm{im}(i \otimes \mathrm{id})$ is an ideal of $B \otimes_{\min} D$ contained in $\ker(p \otimes \mathrm{id})$. Thus $p \otimes \mathrm{id}$ factors through $(B \otimes_{\min} D)/I$ to yield a $*$-homomorphism $\alpha : (B \otimes_{\min} D)/I \to C \otimes_{\min} D$ that will be injective if and only if the tensored sequence is exact. Now the bilinear map

$$(p(b), d) \to (b \otimes d) + I : C \times D \to (B \otimes_{\min} D)/I$$

is well-defined and induces a $*$-homomorphism β on $C \otimes D$ such that $\beta \circ \alpha$ is the identity on the dense subalgebra $(B \otimes D)/I$. Hence exactness comes down to the continuity of β in the spatial norm. But $|||u||| = \|\beta(u)\|$ defines a C^*-norm on $C \otimes D$ for which β is continuous. Therefore, if C or D is nuclear so that $C \otimes D$ has a unique C^*-norm, then the tensored sequence is exact.

7.3 THE MAXIMAL C^*-NORM

For unital C^*-algebras A and B, the tensor product $A \otimes B$ contains canonical copies of A and B, namely $A \otimes 1$ and $1 \otimes B$. Hence any $*$-homomorphism $\rho : A \otimes B \to \mathcal{B}(\mathcal{H})$ has **restrictions**

$$\pi(a) = \rho(a \otimes 1) \quad \text{and} \quad \sigma(b) = \rho(1 \otimes b)$$

to A and B. These are representations that commute ($[\pi(a), \sigma(b)] = 0$ for all a, b) and satisfy

$$\rho\left(\sum a_i \otimes b_i\right) = \sum \pi(a_i)\sigma(b_i) \tag{$*$}$$

In particular, $\|\rho(\sum a_i \otimes b_i)\| \leq \sum \|a_i\| \|b_i\|$. Conversely, if π and σ are commuting representations of A and B on the same space, then $(*)$ defines a representation of $A \otimes B$. With the help of approximate units these facts can be extended to nonunital algebras [Mur90, 6.3.5].

7.3.1

For C^*-algebras A and B and $u \in A \otimes B$, let

$$\|u\|_{\max} = \sup\{\|\rho(u)\| : \rho \text{ is a representation of } A \otimes B\}$$

Then $\|u\|_{\max} < \infty$ from the above, and $\|\cdot\|_{\max}$ is easily seen to be a C^*-norm. Moreover, if $\|\cdot\|_\alpha$ is any C^*-norm on $A \otimes B$ and ρ is a faithful representation of $A \otimes_\alpha B$, then

$$\|u\|_\alpha = \|\rho(u)\| \leq \|u\|_{\max} \qquad \text{for all } u \in A \otimes B$$

Thus $\|\cdot\|_{\max}$ is the **largest C^*-norm** on $A \otimes B$. It follows that $A \otimes B \to A \otimes_\alpha B$ extends to a surjection $A \otimes_{\max} B \to A \otimes_\alpha B$ and hence that C^*-norms on $A \otimes B$ correspond to closed ideals I of $A \otimes_{\max} B$ such that $(A \otimes B) \cap I = \{0\}$. As indicated above, representations of $A \otimes_{\max} B$ correspond to commuting pairs of representations of A and B.

7.3.2

The reason that C^*-norms are necessarily cross-norms can now be seen. If ρ is a faithful representation of $A \otimes_\alpha B$ with restrictions π and σ to A and B, then for all $a \in A$ and $b \in B$,

$$\|a\|\|b\| = \|a \otimes b\|_{\min} \leq \|a \otimes b\|_\alpha = \|\rho(a \otimes b)\| = \|\pi(a)\sigma(b)\| \leq \|a\|\|b\|$$

7.4 CLASSES OF NUCLEAR ALGEBRAS

Recall that a C^*-algebra A is said to be nuclear if for all C^*-algebras B there is exactly one C^*-norm on $A \otimes B$. Equivalently, A is nuclear if $\|\cdot\|_{\min} = \|\cdot\|_{\max}$ on $A \otimes B$ for all B. When A or B is nuclear, it is customary to denote the completion of $A \otimes B$ in its unique C^*-norm by the same notation, $A \otimes B$. We now look briefly at some classes of nuclear algebras, results that are due mainly to Takesaki (see [Tak79]).

7.4.1 Theorem

Commutative C^*-algebras are nuclear.

Outline of the Proof

Let $\|\cdot\|_\alpha$ be a C^*-norm on $A \otimes B$, A commutative. This hypothesis implies that if ρ is an irreducible representation of $A \otimes_\alpha B$, then ρ is unitarily equivalent to a tensor product of irreducible representations of A and B. To see this, let π and σ be the restrictions of ρ. Then

$$\pi(A) \subset \rho(A \otimes_\alpha B)' = \mathbf{C}I$$

so that $\pi = \phi I$ for a multiplicative linear functional ϕ. Hence $\rho(a \otimes b) = \phi(a)\sigma(b)$,

$$\sigma(B)' = \rho(A \otimes_\alpha B)' = \mathbf{C}I$$

and σ is irreducible. Now $\pi_1 = \phi I_1$ is an irreducible representation of A on $\mathcal{H} = \mathbf{C}$, and for all $u = \sum a_i \otimes b_i$ we have

$$(\pi_1 \otimes \sigma)(u) = I_1 \otimes \left(\sum \phi(a_i)\sigma(b_i)\right) = I_1 \otimes \rho(u)$$

so ρ is unitarily equivalent to $\pi_1 \otimes \sigma$. Hence by 7.2.10,

$$\begin{aligned}\|u\|_\alpha &= \sup\{\|\rho(u)\| : \rho \text{ irreducible}\}\\ &\leq \sup\{\|(\pi \otimes \sigma)(u)\| : \pi, \sigma \text{ irreducible}\}\\ &= \|u\|_{\min}\end{aligned}$$

The other inequality follows as in 7.2.11, since the fact that every irreducible representation of $A \otimes_\alpha B$ is of the form $\pi \otimes \sigma$ for $(\pi, \sigma) \in \hat{A} \times \hat{B}$ implies that the set Q defined there is all of $\hat{A} \times \hat{B}$.

7.4.2

Let A be a C^*-algebra and X a locally compact Hausdorff space. Then the bilinear map

$$C_0(X) \times A \to C_0(X, A) : (f, a) \to fa$$

induces a $*$-homomorphism

$$\pi : C_0(X) \otimes A \to C_0(X, A)$$

which extends to an isomorphism of $C_0(X) \otimes_{\min} A$ and $C_0(X, A)$. To see this, note that (by writing elements of $C_0(X) \otimes A$ in the form $\sum f_i \otimes a_i$ with the a_i linearly independent) π is injective, so $\|u\|_\alpha = \|\pi(u)\|$ defines a C^*-norm on $C_0(X) \otimes A$ which, since $C_0(X)$ is nuclear, is the minimal norm. Therefore π extends to the completion in the minimal norm. The image is then closed; since it contains the dense subset consisting of all $\sum f_i a_i$, it must be everything.

7.4.3

A representation $\pi : A \to \mathcal{B}(\mathcal{H})$ is a **factor representation** if the von Neumann algebra generated by $\pi(A)$ is a factor. Although the restrictions π, σ of an irreducible representation ρ of $A \otimes B$ need not be irreducible, it is true that they are factor representations:

$$(\pi(A)'' \cap \pi(A)')' = (\pi(A)' \cup \pi(A))'' \supset (\sigma(B) \cup \pi(A))'' = \rho(A \otimes B)''$$

This observation allows one to argue as in 7.4.1 to prove the following

Theorem

Type I C^*-algebras are nuclear.

Outline of the Proof

If A is type I, ρ is an irreducible representation of $A \otimes_\alpha B$ and π, σ are the restrictions, then $\pi(A)$ generates a type I factor. We have $\sigma(B) \subset \pi(A)'$, and the next step is to show that the von Neumann algebra generated by $\sigma(B)$ is all of $\pi(A)'$, hence a type I factor. Then let E and F be minimal projections in $\pi(A)'$ and $\sigma(B)'$, so that $E\mathcal{H}$ is invariant for $\pi(A)$ and $F\mathcal{H}$ for $\sigma(B)$, and let π_1 and σ_1 be the restrictions of π and σ to $E\mathcal{H}$ and $F\mathcal{H}$. It is easily seen that π_1 and σ_1 are irreducible and that ρ is unitarily equivalent to $\pi_1 \otimes \sigma_1$. Therefore, $\|u\|_\alpha \le \|u\|_{\min}$ for $u \in A \otimes B$ as before.

7.4.4

Let $\mathcal{S}$ be a family of nuclear C^*-subalgebras of a C^*-algebra A such that $\mathcal{S}$ is directed upward by inclusion (i.e., for all $S_1, S_2 \in \mathcal{S}$ there exists $S_3 \in \mathcal{S}$ with $S_1, S_2 \subset S_3$) and has union dense in A. Then A is nuclear: Given a C^*-norm $\|\cdot\|_\alpha$ on $A \otimes B$, $\sum a_i \otimes b_i \in A \otimes B$, and $\varepsilon > 0$, there exist $S \in \mathcal{S}$ and elements $a_i' \in S$ such that $\|a_i - a_i'\| < \varepsilon$, and therefore

$$\begin{aligned}\left\|\sum a_i \otimes b_i\right\|_\alpha &\le \left\|\sum a_i' \otimes b_i\right\|_\alpha + \left\|\sum (a_i - a_i') \otimes b_i\right\|_\alpha \\ &\le \left\|\sum a_i' \otimes b_i\right\|_{\min} + \varepsilon \sum \|b_i\| \\ &\le \left\|\sum a_i \otimes b_i\right\|_{\min} + 2\varepsilon \sum \|b_i\|\end{aligned}$$

If this result is coupled with the general fact that quotients of nuclear algebras are nuclear (more on this later), then we have:

Theorem

Direct limits of nuclear algebras are nuclear.

In particular, AF algebras (hence UHF algebras and the algebra of compact operators) are nuclear. As the UHF algebras show, the class of nuclear C^*-algebras contains the class of type I C^*-algebras properly.

7.5 INJECTIVITY

There is an interesting and important characterization of nuclearity for a C^*-algebra A in terms of the second dual A^{**} (equivalently, by 5.7.2, the enveloping von Neumann algebra A''). A von Neumann algebra $\mathcal{R}$ on a Hilbert space $\mathcal{H}$ is called **injective** if there exists $p : \mathcal{B}(\mathcal{H}) \to \mathcal{R}$ that is linear, acts as the identity on $\mathcal{R}$, and has norm 1. A map with the first two of these properties will be called **retraction** onto $\mathcal{R}$.

7.5.1 ***Theorem*** [ChE77]

A C^*-algebra A is nuclear if and only if A^{**} is injective.

Outline of the Proof

Let A be unital and nuclear. The assertion that A^{**} is injective means that if A is identified with the range of its universal representation so that $A \subset \mathcal{B}(\mathcal{H})$ where $\mathcal{H}$ is the Hilbert space of the universal representation, then there exists a norm 1 retraction $\mathcal{B}(\mathcal{H}) \to A''$. We will consider only the first step in the proof, namely the construction of a retraction $\mathcal{B}(\mathcal{H}) \to A'$ (the existence of a retraction onto A' is equivalent to the existence of a retraction onto A'' [HaT67]). Consider the bilinear map $(a, a') \to aa'$ from $A \times A'$ to $\mathcal{B}(\mathcal{H})$. It induces a $*$-homomorphism $A \otimes A' \to \mathcal{B}(\mathcal{H})$ that extends by continuity to a representation π of $A \otimes_{\max} A'$ on $\mathcal{H}$. Since A is nuclear, we have

$$A \otimes_{\max} A' = A \otimes_{\min} A' \subset A \otimes_{\min} \mathcal{B}(\mathcal{H})$$

by 7.2.12, so by 5.8.3 there is a representation σ of $A \otimes_{\min} \mathcal{B}(\mathcal{H})$ on a Hilbert space $\mathcal{H}' \supset \mathcal{H}$ such that $\mathcal{H}$ is invariant (hence reducing) for $\sigma(A \otimes_{\min} A')$ and

$$\sigma(a \otimes a') \mid_{\mathcal{H}} = \pi(a \otimes a') = aa'$$

for all $a \in A$ and $a' \in A'$. Let P be orthogonal projection of $\mathcal{H}'$ on $\mathcal{H}$ and define $p : \mathcal{B}(\mathcal{H}) \to A'$ by

$$p(x) = P\sigma(1 \otimes x) \mid_{\mathcal{H}} \qquad \text{for all } x \in \mathcal{B}(\mathcal{H})$$

The verification that p is a norm 1 retraction onto A' is straightforward.

7.5.2

It is often necessary to consider *linear* maps between C^*-algebras. Examples (other than homomorphisms) include linear functionals and retractions (see above). Any linear map $\rho : A \to B$ of C^*-algebras induces linear maps $\rho^{(n)} = \rho \otimes \mathrm{id}$ from $A \otimes M_n(\mathbf{C})$ to $B \otimes M_n(\mathbf{C})$, $n = 1, 2, \ldots$. Equivalently, viewing $A \otimes M_n(\mathbf{C})$ as $M_n(A)$ and similarly for B, these maps are given by $\rho^{(n)}(a) = (\rho(a_{ij}))$ for $a = (a_{ij}) \in M_n(A)$. Which properties of ρ are inherited by the $\rho^{(n)}$? If ρ is multiplicative, so is $\rho^{(n)}$, and the same is true of self-adjointness. However, positivity is not necessarily preserved (e.g., the transpose map t on $M_2(\mathbf{C})$ is positive but $t^{(2)}$ is not).

A linear map $\rho : A \to B$ is said to be **completely positive** if the maps $\rho^{(n)}$ are all positive. Lance identified the relevance of this notion for the study of nuclearity [Lan73]. From the above it is clear that $*$-homomorphisms are completely positive. In fact the retraction p constructed in 7.5.1 is completely positive. This is an instance of the general fact that compressing a representation $\pi : A \to \mathcal{B}(\mathcal{H}')$ to a subspace $\mathcal{H} \subset \mathcal{H}'$ yields a completely positive map. Certainly $p = P\pi(\cdot)\mid_{\mathcal{H}}$, being a composition of positive maps, is positive. A short computation shows that $p^{(n)} = P^{(n)}\pi^{(n)}(\cdot)\mid_{\mathcal{H}^n}$, where $P^{(n)}$ is the inflation of P, and so $p^{(n)}$ is positive for the same reason.

Any state $\rho : A \to \mathbf{C}$ is completely positive (hence so is any positive linear functional). This can be seen by direct computation or, since

$$\rho = \langle \pi_\rho(\cdot)\xi_\rho, \xi_\rho \rangle$$

expresses ρ as a compression (to a one-dimensional subspace) of the associated GNS representation, as another instance of the fact that compressions of representations are completely positive. In another direction, Naimark proved that if A is *commutative* and unital, then any unital positive linear map $A \to \mathcal{B}(\mathcal{H})$ is the compression of a representation of A on a larger Hilbert space, hence is completely positive.

These results suggest the question: which completely positive maps arise as compressions of representations?

7.5.3 Theorem [Sti55]

Any unital completely positive map $\phi : A \to \mathcal{B}(\mathcal{H})$ is the compression of a representation of A on a Hilbert space $\mathcal{H}' \supset \mathcal{H}$.

Proof

This is an extension of the GNS construction (the case $\mathcal{H} = \mathbf{C}$), making use of the form defined on the linear space $A \otimes \mathcal{H}$ by

$$((a \otimes \xi), (b \otimes \eta)) \to \langle \phi(b^*a)\xi, \eta \rangle$$

and the action of A on $A \otimes \mathcal{H}$ given by $\pi(a)(b \otimes \xi) = (ab \otimes \xi)$.

7.5.4

A C^*-algebra A is **injective** if, for any C^*-algebras $B \subset B_1$, any completely positive contraction $B \to A$ has a completely positive contractive extension $B_1 \to A$ [EfL77]. In particular for any injective C^*-subalgebra $A \subset \mathcal{B}(\mathcal{H})$ there is a norm 1 retraction $\mathcal{B}(\mathcal{H}) \to A$, the condition considered at the beginning of the section. For von Neumann algebras the two notions are in fact equivalent, so the injectivity of a von Neumann algebra does not depend on the particular space on which it is represented.

7.5.5

Let J be a closed two-sided ideal of a C^*-algebra A. As an application of 7.5.1, we show that A is nuclear if and only if J and A/J are nuclear. Notice first that if

$$0 \to X \to Y \to Z \to 0$$

is an exact sequence of Banach spaces, then by elementary considerations the sequence

$$0 \to Z^* \to Y^* \to X^* \to 0$$

of dual spaces and maps is also exact. In particular,

$$0 \to J^{**} \to A^{**} \to \left(A/J\right)^{**} \to 0$$

is exact. But this particular sequence splits *as a sequence of algebras*. To see this, let σ be a nondegenerate representation of A/J, and apply 5.7.1 to the composition with the quotient map. The result is an σ-weakly continuous map $A^{**} \to \sigma(A/J)''$ with kernel the image of J^{**}. As mentioned in 4.4, it follows that the image of J^{**} is a direct summand of A^{**}. So A^{**} is isomorphic to the direct sum of J^{**} and $(A/J)^{**}$, as algebras. But evidently a direct sum of von Neumann algebras is injective if and only if the summands are.

A "bare-hands" proof of this splitting result can be given by examining, for a positive linear functional ρ on A, the relationship between the GNS representation $(\mathcal{H}_\rho, \pi_\rho)$ for A and the GNS representation $(\mathcal{K}_\rho, \sigma_\rho)$ for J associated to $\rho \mid J$. The inclusion $J \subset A$ induces an isometry of $\mathcal{K}_\rho$ onto a subspace $\mathcal{L}_\rho$ of $\mathcal{H}_\rho$ which is invariant for $\pi_\rho(A)$. Moreover this isometry takes σ_ρ to $\pi_\rho \mid \mathcal{L}_\rho$ on J. Finally, if $\{u_\lambda\}$ is an approximate unit for J, then $\pi_\rho(au_\lambda) \to \pi_\rho(a)$ on $\mathcal{L}_\rho$ in the weak operator topology for all $a \in A$, so $(\pi_\rho(J) \mid \mathcal{L}_\rho)'' = (\pi_\rho(A) \mid \mathcal{L}_\rho)''$. In other words, $\sigma_\rho(J)''$ can be identified as a direct summand of $\pi_\rho(A)''$.

7.5.6

Thus ideals of nuclear algebras are nuclear. What about subalgebras? This is false in general (though true if the subalgebra is **hereditary** in the sense that it contains with any $a \geq 0$ the interval $[0, a]$ of the containing algebra), and so one has the question of which algebras can appear as subalgebras of nuclear algebras. A recent result of E. Kirchberg characterizes them as the C^*-algebras D that are **exact** in the sense that for every exact sequence $0 \to A \to B \to C \to 0$ the tensored sequence

$$0 \to A \otimes_{\min} D \to B \otimes_{\min} D \to C \otimes_{\min} D \to 0$$

of 7.2.13 is exact. So nuclear algebras are exact, and moreover subalgebras of exact algebras are exact. Kirchberg's result [Kir94] is that the Cuntz algebra $\mathcal{O}_2$ is universal among separable exact C^*-algebras, in the sense that the separable exact C^*-algebras are precisely (up to $*$-isomorphism) the C^*-subalgebras of $\mathcal{O}_2$.

7.6 SEMIDISCRETENESS

The "nuclear" terminology was introduced by Lance, following Grothendieck, to suggest approximation by finite rank maps. Call a von Neumann algebra $\mathcal{R}$ **semidiscrete** if there exists a net $\phi_\lambda : \mathcal{R} \to \mathcal{R}$ of finite rank weak*-continuous completely positive unital maps such that $\phi_\lambda(T) \to T$ weak* for all $T \in \mathcal{R}$. It follows that the ϕ_λ^* take the predual $\mathcal{R}_*$ to itself, the starting point for the proof in one direction (see [Lan82]) of the following characterization of nuclearity:

7.6.1 Theorem

A C^*-algebra A is nuclear if and only if A^{**} is semidiscrete.

Combining this result with 7.5.1, we have

$$A^{**} \text{ semidiscrete} \iff A \text{ nuclear} \iff A^{**} \text{ injective}$$

The proof that A is nuclear if A^{**} is semidiscrete depends on an analysis of the state space $S(A \otimes B)$ for arbitrary B. To each $f \in S(A \otimes B)$ corresponds the linear map $T_f : B \to A^*$ given by $(T_f b)(a) = f(a \otimes b)$, and if A^{**} is semidiscrete, then these maps can be approximated by finite rank ones (by composing with the ϕ_λ^*). In 7.5.1 some indication of the proof that A^{**} is injective if A is nuclear was given. In [CE76a; ChE77] the fundamental work of Connes on injective factors (that hyperfiniteness, injectivity and semidiscreteness are equivalent for factors on a separable Hilbert space [Con76]) is used to show that injective von Neumann algebras are semidiscrete, thus completing the proofs of both theorems.

7.6.2

It is natural to ask whether the above approximation property (semidiscreteness) for maps on A^{**} can be replaced by a corresponding property of maps on A itself. A completely positive map $\psi : A \to B$ of C^*-algebras is said to have a (norm) **approximate factorization** if there exists a net

$$\psi_\lambda : A \to M_{n(\lambda)}(\mathbf{C}) \to B$$

of compositions of completely positive maps such that $\psi_\lambda(a) \to \psi(a)$ in norm for all $a \in A$. If the identity map on A has an approximate factorization, then it it almost obvious that A^{**} is semidiscrete and hence that A is nuclear. In fact this property is equivalent to nuclearity [Lan82].

7.6.3

There is an interesting application of this characterization of nuclearity to the study of C^*-algebra extensions

$$0 \to \mathcal{K} \to E \to A \to 0$$

of A by $\mathcal{K} = \mathcal{K}(\mathcal{H})$, the algebra of compact operators on the Hilbert space $\mathcal{H}$. It can be assumed that $\mathcal{K}$ is mapped to an essential ideal of E (extensions for which this is not true can be studied as extensions of an appropriate quotient of A). The corresponding (by 3.12.6) $*$-homomorphisms $A \to \mathcal{C}(\mathcal{H})$, the Calkin algebra of $\mathcal{H}$, are precisely the injective ones. It was indicated in 3.11.5 that for $A = C(\partial\mathbf{D})$ the equivalence classes of such extensions form in a natural way a group isomorphic to $\mathbf{Z}$. The operation defined there makes sense for any A, and one can ask if the resulting abelian semigroup Ext A is necessarily a group [BDF77]. Arveson pointed out the importance of complete positivity in this question, leading to the affirmative solution by Choi and Effros in the case that A is nuclear [CE76b]. It is hardly possible to improve on Arveson's account of this [Arv77], but we will give a brief description.

To begin with, the neutral element of Ext A is obtained by taking any faithful representation $\pi : A \to \mathcal{B}(\mathcal{H})$ such that $A \cap \mathcal{K}(\mathcal{H}) = \{0\}$. For example, the direct sum of countably many copies of any faithful representations has this property. Then the C^*-subalgebra of $\mathcal{B}(\mathcal{H})$ generated by A and $\mathcal{K}(\mathcal{H})$ contains $\mathcal{K}(\mathcal{H})$ as an ideal with quotient $\pi(A) \cong A$. It is far from obvious that any two such extensions are equivalent, or that this class functions as the neutral element, but this is known to be true when A is separable.

Now consider any decomposition $\mathcal{H} = \mathcal{H}_1 \oplus \mathcal{H}_2$ into infinite-dimensional subspaces such that in the corresponding 2×2 matrix representation of the operators $\pi(a)$, $a \in A$, the off-diagonal entries are compact. Then the compressions ρ_i to the $\mathcal{H}_i$ are completely positive maps (by 7.5.2) which when composed with the quotient maps $\mathcal{B}(\mathcal{H}_i) \to \mathcal{C}(\mathcal{H}_i)$ are in fact $*$-homomorphisms that correspond to a pair of extensions inverse to one another.

Arveson's idea is now clear: given an extension, if the corresponding $*$-homomorphism $A \to \mathcal{C}(\mathcal{H})$ has a completely positive lifting to $A \to \mathcal{B}(\mathcal{H})$, then Stinespring's theorem 7.5.3 will provide a dilation to a representation on a larger space, hence an inverse extension by proceeding as above. So if for C^*-algebras A and B we denote by $CP(A, B)$ the set of completely positive maps from A to B, and if J is an ideal of B, we would like to know when the map $CP(A, B) \to CP(A, B/J)$, obtained by composing with the quotient map $B \to B/J$, is surjective. Arveson shows the following

i. If A is separable, then $CP(A, B) \to CP(A, B/J)$ has closed range.

ii. If $A = M_n(\mathbf{C})$, then it is surjective.

It follows that every composition

$$A \to M_n(\mathbf{C}) \to B/J$$

of completely positive maps lifts and hence by 7.6.2, if A is nuclear and separable, that every completely positive map lifts.

7.7 AMENABILITY

In this section we will examine amenability in operator algebras. The idea of amenability arose in the investigation, by Polish mathematicians in the 1920s and 1930s, of the existence of an invariant mean for the action of a group G on a set X. In his work on the Banach-Tarski paradox, von Neumann calls a group amenable if the action of the group on itself has an invariant mean. Precisely, any group G acts on the C^*-algebra $\ell^\infty(G)$ by "left translation":

$$(\lambda_y f)(x) = f(y^{-1}x) \qquad \text{for } f \in \ell^\infty(G) \text{ and } x, y \in G$$

A (left) **invariant mean** on G is then a state μ of $\ell^\infty(G)$ such that $\mu(\lambda_y f) = \mu(f)$ for all $f \in \ell^\infty(G)$ and $y \in G$, and (in a particularly pleasant piece of terminology) G is called **amenable** if it has an invariant mean. Here G has been regarded as a discrete group, but the definition extends to locally compact topological groups by substituting $L^\infty(G)$ for $\ell^\infty(G)$. Many common groups are amenable (e.g., any abelian group), and the class is stable under many operations (e.g., subgroups of discrete groups). An example of a nonamenable group is F_2, the free (nonabelian) group on two generators (due to von Neumann). For more information about amenability, a good recent source is [Pat88].

7.7.1

Let A be a Banach algebra. A two-sided module X over A is called a **Banach A-module** if it is a Banach space and there is a constant K such that

$$\|ax\|, \|xa\| \leq K\|a\|\|x\|$$

for all $a \in A$ and $x \in X$. For example, if J is a closed two-sided ideal of A, then both J and A/J are naturally Banach A-modules, as are finite direct sums of Banach A-modules. More important for present purposes, if X is a Banach A-module, then so is the dual space X^* with respect to the induced actions

$$(af)(x) = f(xa) \quad \text{and} \quad (fa)(x) = f(ax)$$

for all $a \in X$, $x \in X$, and $f \in X^*$. The usual notion of a **derivation** on an algebra A, that is, a linear map $D : A \to A$ such that

$$D(ab) = D(a)b + aD(b) \qquad \text{for all } a, b \in A$$

extends without change to maps D from A to a bimodule over A. If X is a Banach A-module, then to each $x \in X$ corresponds a bounded derivation $D_x : A \to X$ given by $D_x a = ax - xa$; these are the **inner derivations**. Finally, A is said to be **amenable** if every bounded derivation from A to a dual Banach A-module X^* (as above) is inner.

The connection with group amenability is provided by the following result of B. E. Johnson [Joh72], who introduced and studied Banach algebra amenability: a locally compact group G is amenable if and only if the convolution algebra $L^1(G)$ is an amenable Banach algebra.

7.7.2

A C^*-algebra is **amenable** if it is amenable as a Banach algebra. There is a C^*-algebra analogue of Johnson's theorem, at least for discrete groups G: The group C^*-algebra $C^*(G)$ is amenable if and only if G is amenable. Group C^*-algebras were introduced briefly in 3.6 and will be considered in more detail in the next chapter. Paterson, building on a corresponding result of de la Harpe for von Neumann algebras, has recently found another connection with group amenability: a C^*algebra A is amenable if and only if its unitary group $G = U(A)$ is "amenable" in the sense that a certain subspace of $\ell^\infty(G)$ (the right uniformly continuous functions) admits a left invariant mean [Pat92].

7.7.3

In the von Neumann algebra context the class of modules taken into consideration needs to be restricted. If $\mathcal{R}$ is a von Neumann algebra, then a dual $\mathcal{R}$-module X^* is said to be **normal** if the maps $a \to af, fa$ $(a \in \mathcal{R},\ f \in X^*)$ are continuous relative to the weak* topologies, and then $\mathcal{R}$ is **amenable** if every bounded derivation from $\mathcal{R}$ to a normal dual $\mathcal{R}$-module is inner. Note that in this definition the derivation was not required to be weak* continuous; in fact this is automatic.

There is a close connection with the C^*-algebra case: if A is an amenable C^*-algebra, then A^{**} is amenable as a von Neumann algebra. Given the automatic weak* continuity of derivations, this is almost obvious.

7.7.4

Recall (from 4.10) that a Neumann algebra $\mathcal{R}$ is hyperfinite if it contains an increasing sequence of finite dimensional subalgebras whose union is σ-weakly dense in $\mathcal{R}$. Prior to the "injective factors paper" of Connes [Con76], it was known that for von Neumann algebras hyperfinite implies injective and that semidiscrete implies injective. Connes showed that all three are equivalent for factors with separable predual. Because these properties behave well with respect to direct integral decomposition into factors (see 4.6.3), it follows that they are equivalent for arbitrary von Neumann algebras with separable predual. In [Con78] it is shown that amenability can be added to this list. Thus for von

Neumann algebras $\mathcal{R}$ with separable predual,

$$\mathcal{R} \text{ hyperfinite} \iff \mathcal{R} \text{ injective} \iff \mathcal{R} \text{ semidiscrete} \iff \mathcal{R} \text{ amenable}$$

There is perhaps no completely satisfactory term to describe this class; Connes has suggested that "amenable" might be the best. For a good account of these developments, see Connes's survey paper [Con82] or his book [Con94, ch.5].

The injective factors paper does more: It gives a virtually complete classification of injective factors (since completed by Haagerup) and hence of amenable von Neumann algebras with separable predual.

For arbitrary C^*-algebras A we have

$$A \text{ amenable} \iff A^{**} \text{ amenable} \iff A \text{ nuclear}$$

Thus nuclearity is characterized by the property that the von Neumann algebras generated by all GNS representations are amenable. The proof that nuclear algebras are amenable (as C^*-algebras) is due to Haagerup [Haa83]. In view of the von Neumann algebra results, it is natural to wonder whether separable nuclear C^*-algebras might be classifiable [Ell93].

Chapter Eight

Crossed Products

8.1 C^*-DYNAMICAL SYSTEMS

The idea of a C^*-dynamical system (G, α, A) was introduced in 3.8. Here G is a locally compact Hausdorff topological group, A is a C^*-algebra and α is an action of G on A (i.e., a homomorphism $s \to \alpha_s$ from G into Aut A, the group of automorphisms of A, such that $s \to \alpha_s(a)$ is continuous for all $a \in A$). The original impetus for studying such systems comes from **topological dynamics**, where $A = C_0(X)$ with X locally compact and Hausdorff and α is induced by an action $(s, x) \to s^{-1}x$ of G on X by homeomorphisms:

$$(\alpha_s(f))(x) = f(s^{-1}x) \qquad \text{for } f \in C_0(X)$$

Consideration of the representation theory of C^*-dynamical systems leads immediately to the notion of crossed product C^*-algebras and thus opens the door to the application of C^*-algebra techniques in the study of topological dynamics. There is an entirely parallel development connecting ergodic theory, W^*-dynamical systems and W^*-crossed products (4.9). G.K. Pedersen's book [Ped79] is a good general reference for this topic.

8.2 COVARIANT REPRESENTATIONS

A **(covariant) representation** of a C^*-dynamical system (G, α, A) is a pair (π, U) where π is a nondegenerate representation of A on a Hilbert space $\mathcal{H}$ and U is a unitary representation (3.6.2; 5.9) of G on the same space such that

$$\pi(\alpha_s(a)) = U_s \pi(a) U_s^* \qquad \text{for all } a \in A, \quad s \in G$$

Thus the action of G leaves $\ker \pi$ invariant and the induced action on $\operatorname{im} \pi$ is the restriction of $\operatorname{Ad} U$. Covariant representations (π, U) on $\mathcal{H}$ and (π', U') on $\mathcal{H}'$ of (G, α, A) are **unitarily equivalent** if there is a unitary operator $W : \mathcal{H} \to \mathcal{H}'$ such that

$$\pi'(a) = W\pi(a)W^* \quad \text{and} \quad U'_t = WU_tW^*$$

for all $a \in A$ and $t \in G$.

8.2.1

To construct covariant representations we need the following fundamental fact about locally compact topological groups: Any such group G admits a unique (up to scalar multiples) nonzero left invariant Borel measure, called the **left Haar measure** of G. Using it, one can define, for any C^*-algebra A, the convolution product on the linear space $C_c(G, A)$ of compactly supported continuous functions from G to A:

$$(f * g)(t) = \int_G f(s)g(s^{-1}t)ds$$

for all $f, g \in C_c(G, A)$ and $t \in G$, where ds denotes Haar measure. Moreover we have a norm

$$\|f\|_1 = \int_G \|f(s)\|ds$$

on $C_c(G, A)$ that satisfies

$$\|f * g\|_1 \leq \|f\|_1 \|g\|_1 \qquad \text{for all } f, g \in C_c(G, A)$$

There is also an involution on $C_c(G, A)$, but to define this, we have to take into account the fact that left Haar measure need not be right invariant. However, from the uniqueness of Haar measure, it follows that there is a continuous homomorphism

$$\Delta : G \to (0, +\infty)$$

called the **modular function**, such that

$$\int_G \Delta(s^{-1})g(s^{-1})ds = \int_G g(s)ds \qquad \text{for all } g \in C_c(G)$$

If G is abelian, compact, or discrete, then left Haar measure is right invariant so $\Delta = 1$ (G is **unimodular**). For $f \in C_c(G, A)$ and $t \in G$, let

$$f^*(t) = \Delta(t^{-1})f(t^{-1})^*$$

This defines an involution on $C_c(G, A)$. It would be an involution even without Δ, but with it the involution is isometric:

$$\|f^*\|_1 = \int_G \|f^*(s)\|ds = \int_G \Delta(s^{-1})\|f(s^{-1})\|ds = \int_G \|f(s)\|ds = \|f\|_1$$

The reader should note that other conventions are in use, such as using right Haar measure or incorporating the modular function in the norm instead of in the involution.

8.2.2

The above considerations apply in particular to the case of a C^*-dynamical system (G, α, A). Note, however, that they take no account of the action of G on A. It is a rather remarkable fact that the operations on $C_c(G, A)$ can be modified so that nondegenerate $*$-algebra representations correspond bijectively to the covariant representations of (G, α, A). Namely for all $f, g \in C_c(G, A)$ and $t \in G$ define

$$(f * g)(t) = \int_G f(s)\alpha_s(g(s^{-1}t))ds$$

$$f^*(t) = \Delta(t^{-1})\alpha_t(f(t^{-1})^*)$$

With these operations and the norm defined in 8.2.1, we again have a normed algebra with an isometric involution. The completion $L^1(G, \alpha, A)$ is therefore a Banach $*$-algebra (but not a C^*-algebra, since the L^1-norm is a C^*-norm only in trivial cases).

8.2.3

The correspondence between covariant representations of (G, α, A) and $*$-representations of $L^1(G, \alpha, A)$ will now be described. Given a covariant representation (π, U) of (G, α, A) on $\mathcal{H}$, define a representation $\pi \times U$ of $C_c(G, A)$ on $\mathcal{H}$ by

$$(\pi \times U)(f) = \int_G \pi(f(s))U_s ds$$

Here the integral may be understood in the weak sense:

$$\langle((\pi \times U)f)\xi, \eta\rangle = \int_G \langle\pi(f(s))U_s\xi, \eta\rangle ds$$

for all $\xi, \eta \in \mathcal{H}$. It is straightforward to check that this equation defines an operator $(\pi \times U)(f)$ on $\mathcal{H}$ of norm at most $\|f\|_1$ and that $\pi \times U$ is a $*$-representation of $C_c(G, A)$. It is nondegenerate (because π is) and it extends by continuity to a $*$-representation of $L^1(G, \alpha, A)$.

Conversely, suppose that a nondegenerate representation ρ of $L^1(G, \alpha, A)$ on $\mathcal{H}$ is given. In the special case where G is discrete and A unital, it is easy to see that ρ has the form $\pi \times U$. Just as in 3.8.3, $C_c(G, A)$ contains canonical copies of A and U, and we get the desired covariant representation (π, U) by restricting ρ to these copies. In general, take

$$\pi(a)\rho(f)\xi = \rho(af)\xi \quad \text{and} \quad U_t\rho(f)\xi = \rho(\alpha_t \circ f_t)\xi$$

for $a \in A$, $t \in G$ and $f \in C_c(G, A)$, where $f_t(s) = f(t^{-1}s)$. These can be shown to be well-defined and bounded by means of an argument using an approximate unit for $C_c(G, A)$. For the approximate unit take $e_\lambda = u_\lambda\phi_\lambda$ with $\{u_\lambda\}$ and $\{\phi_\lambda\}$ approximate units in A and $C_c(G)$. The essential idea is that because ρ is nondegenerate, it can be extended uniquely to the multiplier algebra of $C_c(G, A)$ (see 5.8.1), and the latter always contains canonical copies of G and A.

In fact Raeburn has used this idea as the basis for an alternative approach to the crossed product, namely as a C^*-algebra B for which there are morphisms $A \to M(B)$ and $G \to \mathcal{U}(M(B))$ with suitable properties [Rae88].

It is not difficult to verify that this bijection respects unitary equivalence.

8.2.4

How do we know that a C^*-dynamical system (G, α, A) has any covariant representations? To any representation π of A on a Hilbert space $\mathcal{H}$ (and by the Gelfand-Naimark theorem (5.4.1) we know that there are plenty of these) is associated in a natural way—via the left regular representation λ of G—a covariant representation $(\tilde{\pi}, \tilde{\lambda})$ of (G, α, A) on $L^2(G, \mathcal{H})$: For all $a \in A$, $s, t \in G$, and $\xi \in L^2(G, \mathcal{H})$, let

$$(\tilde{\pi}(a)\xi)(t) = \pi(\alpha_{t^{-1}}(a))\xi(t) \quad \text{and} \quad (\tilde{\lambda}_s\xi)(t) = \xi(s^{-1}t)$$

Notice that $\tilde{\lambda}$ is just λ taken with multiplicity $\dim \mathcal{H}$. The covariance condition is easy to check, and one has, for the corresponding representation of

$L^1(G, \alpha, A)$,

$$(((\tilde{\pi} \times \tilde{\lambda})f)\xi)(t) = \int_G \pi(\alpha_{t^{-1}}(f(s)))\xi(s^{-1}t)ds$$

for all $f \in C_c(G, A)$, $\xi \in L^2(G, \mathcal{H})$, and $t \in G$. If π is nondegenerate, so are $\tilde{\pi}$ and $\tilde{\pi} \times \tilde{\lambda}$. We will refer to $\tilde{\pi} \times \tilde{\lambda}$ (and $(\tilde{\pi}, \tilde{\lambda})$) as the **regular representation associated to π.**

8.2.5

Another type of covariant representation is associated to any action $G \times X \to X$ for which there is a G-invariant Borel measure μ on X. We then have a C^*-dynamical system $(G, \alpha, C_0(X))$ for which the representations (M, λ) on $L^2(X, \mu)$ given by

$$M(f)\xi = f\xi \qquad \text{for } f \in C_0(X),\ \xi \in L^2(X, \mu)$$

$$(\lambda_s\xi)(t) = \xi(s^{-1}t) \qquad \text{for } s, t \in G,\ \xi \in L^2(X, \mu)$$

form a covariant pair.

A point of interest here is that if G acts ergodically on X (see 4.9.3), then $M \times \lambda$ is irreducible. This is proved by showing that the image has trivial commutant. But if T is in the commutant, it commutes with all $M(f)$, $f \in C_0(X)$, hence has the form $M(\phi)$ for some $\phi \in L^\infty(X)$. Since it also commutes with all λ_s, $T \in \mathbf{C}I$ follows by ergodicity.

8.3 CROSSED PRODUCTS

Let (G, α, A) be a C^*-dynamical system. The (full) **C^*-crossed product**, denoted $A \rtimes_\alpha G$, is the completion of $C_c(G, A)$ (or of $L^1(G, \alpha, A)$) in its greatest C^*-norm $\|f\| = \sup \|\pi(f)\|$, the supremum being taken over all $*$-representations π of $L^1(G, \alpha, A)$. Thus, by 8.2.3, the $*$-representations of $A \rtimes_\alpha G$ correspond bijectively to the covariant representations of (G, α, A). The **reduced C^*-crossed product** $A \rtimes_{\alpha r} G$ is the norm closure of the image of $\tilde{\pi}_S \times \tilde{\lambda}$, where π_S is the universal representation (5.7) of A.

8.3.1

From the definition it follows that $C_c(G, A)$ is dense in $A \rtimes_\alpha G$ and that any $*$-representation of $C_c(G, A)$ extends to a representation of $A \rtimes_\alpha G$. In particular,

$A \rtimes_{\alpha r} G$ is a quotient of $A \rtimes_{\alpha} G$. Likewise $C_c(G, A)$ is dense in $A \rtimes_{\alpha r} G$ and, for any representation π of A, $\tilde{\pi} \times \tilde{\lambda}$ extends to a representation of $A \rtimes_{\alpha r} G$. It can be shown that if π is faithful, so is $\tilde{\pi} \times \tilde{\lambda}$ [Ped 79, 7.7.5]; hence the reduced crossed product is given (up to $*$-isomorphism) by the closure of the image of $\tilde{\pi} \times \tilde{\lambda}$ for any faithful representation π of A.

8.3.2

Here are some special cases. If $A = \mathbf{C}$, the nondegenerate representations ρ of $C_c(G)$ correspond to unitary representations U of G via the formula

$$\rho(f) = \int_G f(s) U_s ds \qquad \text{for all } f \in C_c(G)$$

The completion of $C_c(G)$ in its greatest C^*-norm is called the **group C^*-algebra** of G, denoted $C^*(G)$. Thus

$$C^*(G) = \mathbf{C} \rtimes_{\text{triv}} G$$

Likewise the **reduced group C^*-algebra** of G, denoted $C_r^*(G)$, is obtained by taking $A = \mathbf{C}$ in the definition of the reduced crossed product. Thus it is the norm closure of the regular representation λ of $C_c(G)$ on $L^2(G)$ associated to the one-dimensional identity representation π of $A = \mathbf{C}$ (see 8.2.4), namely

$$(\lambda(f)\xi)(t) = \int_G f(s)\xi(s^{-1}t) ds$$

for $f \in C_c(G)$, $\xi \in L^2(G)$, and $t \in G$.

The map $\sum f_i \otimes a_i \to \sum f_i a_i : C_c(G) \otimes A \to C_c(G, A)$ is always linear and injective. If the action of G on A is trivial, then as noted in 3.8.5 this map is even a $*$-homorphism, and it follows that

$$A \rtimes_{\text{triv}} G \cong C^*(G) \otimes_{\max} A$$

More generally, the same is true if the action of G is inner, in the sense that there is a homomorphism U from G to the unitary group of the multiplier algebra $M(A)$ such that $\alpha = \text{Ad } U$.

If G is a semidirect product $N \rtimes H$, then

$$C^*(G) \cong C^*(N) \rtimes_{\alpha} H$$

The hypothesis means that G is generated by the subgroups N and H with N normal and $N \cap H = \{e\}$. The action of H on N, given by conjugation, induces the action α of H on $C_c(N)$, and hence on $C^*(N)$.

8.3.3

C^*-dynamical systems (G, α, A) and (G, β, B) are **covariantly isomorphic** if there is a $*$-isomorphism $A \to B$ that respects the actions of G. It is clear that in this case there is a $*$-isomorphism $A \rtimes_\alpha G \to B \rtimes_\beta G$ and similarly for the reduced crossed product.

8.3.4

The usefulness of of group C^*-algebras and crossed products has naturally created a good deal of interest in extending the ideas to more general settings. We have already looked at the case of groupoids (in 3.13). Some recent results are concerned with the situation in which a semigroup acts on a C^*-algebra by automorphisms or even by endomorphisms [Mur94; ALN94]. One somewhat unexpected feature is that the C^*-algebra of an commutative semigroup need not be commutative. Indeed the case $C^*(\mathbf{N})$ already is interesting: it is the Toeplitz algebra (see 3.11).

8.4 AMENABLE GROUPS

Recall (from 7.7) that a locally compact group G is amenable if it admits a left invariant mean, that is, a positive linear functional m on $L^\infty(G)$ such that $m(1) = 1$ and

$$m(\lambda_s f) = m(f) \qquad \text{for all } f \in L^\infty(G) \text{ and } s \in G$$

where $(\lambda_s f)(t) = f(s^{-1}t)$. When G is amenable, its group C^*-algebra $C^*(G)$ is a much more accessible object than it is in general. In fact G is amenable if and only if the quotient map $C^*(G) \to C^*_r(G)$ is injective. More generally, it is true that

$$A \rtimes_\alpha G \cong A \rtimes_{\alpha r} G$$

whenever G is amenable. These results are somewhat difficult, and we refer to [Ped79, ch. 7] for the details. There is also a connection with nuclearity. If G is amenable, then $C^*(G)$ is nuclear; the converse is true for discrete groups.

If (G, α, A) is a C^*-dynamical system with G amenable and A nuclear, then $A \rtimes_\alpha G$ is nuclear [Goo82].

8.4.1

Amenable groups give rise to a general class of simple C^*-algebras. An action $G \times X \to X$ of a locally compact group on a locally compact space is **topologically free** if $gx = x$ for some $g \in G$ and $x \in X$ is only possible when g is the identity element. The action is **minimal** if every orbit $O_x = \{gx : g \in G\}$ is dense in X.

Theorem

Let (G, α, X) be a topological dynamical system. If G is amenable and acts freely, then $C_0(X) \rtimes_\alpha G$ is simple if and only if the action is minimal.

The measure-theoretic (i.e., von Neumann algebra) version of this theorem has already been considered in connection with the construction of factors (4.9.3); see [Goo82] for a discussion of both these results.

An important example is the irrational rotation algebra $A_\theta = C(S^1) \rtimes_\theta \mathbf{Z}$, described in 3.9. The irrationality of θ implies that the action is topologically free and minimal; hence A_θ is simple.

8.4.2

What about rational rotations? Let $\mathbf{Z}_n$ be the cyclic group of order n, acting on S^1 by rotation through (multiples of) $2\pi/n$, and consider the algebra $A = C(S^1) \rtimes \mathbf{Z}_n$. The full and reduced crossed products are the same, so by 8.3.1 we can identify A with its image under $\tilde{\pi} \times \tilde{\lambda}$ for any faithful representation π of $C(S^1)$. Take π to be the multiplication representation M on $\mathcal{H} = L^2(S^1)$, and note that $L^2(G, \mathcal{H}) = L^2(\mathbf{Z}_n, L^2(S^1))$ is a direct sum of n copies of $L^2(S^1)$. Since $C(S^1)$ is generated by the single (unitary) element z, we see that the image of $\tilde{\pi} \times \tilde{\lambda}$ is generated by unitaries u (the image of z) and v (the image of the generator of $\mathbf{Z}_n$). Moreover taking $f(z) = z$ in the covariance condition $vM(f)v^* = \alpha(M(f))$ gives $vuv^* = \omega u$, where $\omega = \exp(-2\pi i/n)$. In fact it is easy to see that u and v are given by the $n \times n$ operator matrices

$$u = \begin{pmatrix} z & & & 0 \\ & \overline{\omega} z & & \\ & & \ddots & \\ 0 & & & \overline{\omega}^{n-1} z \end{pmatrix}, \quad v = \begin{pmatrix} 0 & & & 1 \\ 1 & \ddots & & \\ & \ddots & \ddots & \\ 0 & & 1 & 0 \end{pmatrix}$$

but we will not need this. Instead we can proceed as follows:

i. Show that $e = \sum_{j=1}^{n} v^j/n$ is a projection in A.
ii. Show that $eAe \cong C(S^1)$.
iii. Show that $e_{ij} = u^i e u^{-j}$ $(0 \leq i, j \leq n-1)$ is a full set of matrix units in A; namely, $e_{ij}e_{kl} = \delta_{jk}e_{il}$ and $\sum_{i=1}^{n} e_{ii} = 1$.

Together these imply that $A \cong M_n(C(S^1))$.

8.4.3

Let G be amenable and act on itself by left translation $\gamma : (s, t) \to s^{-1}t$. This is free and minimal, and so $C_0(G) \rtimes_\gamma G$ is simple by 8.4.1. We can make use of 8.2.5 to determine what this algebra is. Notice that there is a G-invariant measure on $G = X$, namely left Haar measure, and that the action is ergodic for this measure. Hence the representation $M \times \lambda$ of $C_0(G) \rtimes G$ on $L^2(G)$ described in 8.2.5 is irreducible. Since the crossed product is simple, this representation is also faithful; thus $C_0(G) \rtimes G$ can be identified with its image in $\mathcal{B}(L^2(G))$.

Theorem

If G is amenable, then $C_0(G) \rtimes_\gamma G \cong \mathcal{K}(L^2(G))$.

Proof

It is enough to show that the image of $M \times \lambda$ is $\mathcal{K}(L^2(G))$, and for this we only need to show that it intersects $\mathcal{K}(L^2(G))$ nontrivially (for then, by irreducibility and 3.11.2, it must contain $\mathcal{K}(L^2(G))$, hence be equal to it by simplicity). By 8.2.3

$$((M \times \lambda)h)\xi = \int_G h(s)\lambda_s \xi \, ds$$

for all $h \in C_c(G, C_0(G))$ and $\xi \in L^2(G)$. Fix nonzero $f, g \in C_c(G)$, and let

$$(h(s))(t) = f(s^{-1}t)g(t)\Delta(s^{-1}t)$$

where Δ is the modular function of G. Then

$$\begin{aligned}(((M \times \lambda)h)\xi)(t) &= \int_G f(s^{-1}t)g(t)\Delta(s^{-1}t)\xi(s^{-1}t)ds \\ &= g(t)\int_G f(s^{-1})\xi(s^{-1})\Delta(s^{-1})ds \\ &= g(t)\int_G f(s)\xi(s)ds \\ &= g(t)\langle \xi, \overline{f}\rangle\end{aligned}$$

so that $(M \times \lambda)h$ is an operator on $L^2(G)$ of rank one.

8.4.4

The Bunce-Deddens algebras, described in 3.10.3, can be identified as crossed product algebras. There is such an algebra for each sequence $1 \leq n_1 < n_2 < \cdots$ of integers such that each divides the next, but to simplify the notation we will consider only the Bunce-Deddens algebra of type 2^∞, namely $A = \lim_\rightarrow M_{2^n}(\mathbf{C}) \otimes C(S^1)$, where the first map

$$\phi : C(S^1) \rightarrow M_2(\mathbf{C}) \otimes C(S^1) = C(S^1, M_2(\mathbf{C}))$$

is the "twice around" embedding of multiplicity 2,

$$(\phi(f))(z) = \begin{pmatrix} f(z^2) & 0 \\ 0 & f(z^2) \end{pmatrix}$$

and subsequent maps are of the form id $\otimes\phi$.

Let $\mathbf{Z}[\frac{1}{2}]$ be the group of dyadic rational numbers, and let D be the quotient group $\mathbf{Z}[\frac{1}{2}]/\mathbf{Z}$. The action $(d, z) \rightarrow e^{2\pi i d}z$ of D on S^1 is free and minimal, and so by 8.4.1 the crossed product $C(S^1) \rtimes D$ is simple. To identify this algebra as the Bunce-Deddens algebra of type 2^∞, observe that D is the direct limit of the system

$$\mathbf{Z}_2 \rightarrow \mathbf{Z}_4 \rightarrow \mathbf{Z}_8 \rightarrow \cdots$$

where the maps are all multiplication by 2. It follows that $C(S^1) \rtimes D$ is the direct limit of the induced system

$$C(S^1) \rtimes \mathbf{Z}_2 \to C(S^1) \rtimes \mathbf{Z}_4 \to \cdots$$

Applying 8.4.2 then gives the result. We are also making use of the fact that the action $D \to \operatorname{Aut} C(S^1)$ is the direct limit of the actions $\mathbf{Z}_{2^n} \to \operatorname{Aut} C(S^1)$ of 8.4.2.

8.5 CONTINUOUS FIELDS

The concept of a continuous field of C^*-algebras over a topological space was mentioned briefly in 3.5. It was noted there that to any continuous field is associated in a canonical way a C^*-algebra (consisting of certain sections of the field). The following question naturally presents itself: For a given C^*-algebra A and space X, when is there a field over X for which the associated C^*-algebra is A? When X is locally compact and Hausdorff there is a very satisfactory answer (see [ENN93] for a discussion of the proof).

8.5.1 Theorem [Lee76]

Let A be a C^*-algebra and X a locally compact Hausdorff space. Then A arises from a continuous field of C^*-algebras over X if and only X is a continuous open image of Prim(A), *the primitive ideal space of A (6.2). In this case the continuous field is obtained by assigning to each $x \in X$ the quotient algebra $A_x = A/I_x$, where I_x is the intersection of the primitive ideals of A in the preimage of x.*

In this section we are interested in the interplay between continuous fields and crossed products. Specifically, if A arises from a continuous field over a space X and a group G acts on A, when does $A \rtimes G$ arise from a continuous field over the same space X?

8.5.2 Theorem [Wil89]

Let (G, α, A) be a C^*-dynamical system with G amenable, and assume that A arises from a continuous field of C^*-algebras $\{A_x : x \in X\}$ over a locally compact Hausdorff space X. If $A_x = A/I_x$ as in 8.5.1 and $\alpha_g(I_x) \subset I_x$ for all $g \in G$ and $x \in X$, then $A \rtimes_\alpha G$ arises from a continuous field over X with fibers $(A \rtimes_\alpha G)_x = A_x \rtimes_{\alpha^x} G$, where α^x is the induced action of G on A_x.

8.5.3

There is an interesting example with $A = C_0(\mathbf{R}^2)$. Since it is commutative, the primitive ideal space coincides with the maximal ideal space $\mathbf{R}^2$. The projections of $\mathbf{R}^2$ on the coordinate spaces are continuous and open, and so (8.5.1) A can be regarded as arising from a continuous field over $\mathbf{R}$. To be specific take the projection $(x, y) \to x$ so that

$$I_x = \{f \in C_0(\mathbf{R}^2) : f(x, y) = 0 \text{ for all } y\}$$

This is the kernel of the map $\phi_x : C_0(\mathbf{R}^2) \to C_0(\mathbf{R})$ defined by $(\phi_x(f))(y) = f(x, y)$, and so

$$A_x = A/I_x \cong C_0(\mathbf{R})$$

for all $x \in \mathbf{R}$.

Now consider the action of $G = \mathbf{R}$ on A given by

$$(\alpha_t(f))(x, y) = f(x, tx + y)$$

for all $f \in C_0(\mathbf{R}^2)$ and $x, y, t \in \mathbf{R}$. The ideals I_x are invariant for this action, and so by 8.5.2 the crossed product $C_0(\mathbf{R}^2) \rtimes_\alpha \mathbf{R}$ arises from a continuous field of C^*-algebras over $\mathbf{R}$ with fibers $C_0(\mathbf{R}) \rtimes_{\alpha^x} \mathbf{R}$. The action α^x of $\mathbf{R}$ on $A_x = C_0(\mathbf{R})$ is given by

$$(\alpha_t^x(f))(y) = f(tx + y)$$

for $f \in C_0(\mathbf{R})$ and $x, y, t \in \mathbf{R}$. When $x = 0$, this is the trivial action, and therefore by 8.3.2,

$$(C_0(\mathbf{R}^2) \rtimes_\alpha \mathbf{R})_0 \cong C_0(\mathbf{R}) \rtimes_{\alpha^0} \mathbf{R} \cong C_0(\mathbf{R}) \otimes C^*(\mathbf{R})$$

Since $C_0(\mathbf{R})$ is nuclear (7.4.1), it is not necessary to specify which tensor product is intended here. It was noted in 3.6.2 that $C^*(\mathbf{R}) \cong C_0(\hat{\mathbf{R}})$, and since $\hat{\mathbf{R}} \cong \mathbf{R}$ (see below, 8.6.1), we have by 7.4.2 that

$$(C_0(\mathbf{R}^2) \rtimes_\alpha \mathbf{R})_0 \cong C_0(\mathbf{R}) \otimes C_0(\mathbf{R}) \cong C_0(\mathbf{R}^2)$$

On the other hand, when $x \neq 0$, the action is (essentially) given by the action of $\mathbf{R}$ on itself by translation, and thus by 8.4.3 we have

$$(C_0(\mathbf{R}^2) \rtimes_\alpha \mathbf{R})_x \cong C_0(\mathbf{R}) \rtimes_{\alpha^x} \mathbf{R} \cong \mathcal{K}(L^2(\mathbf{R}))$$

In other words, the field of C^*-algebras associated to $C_0(\mathbf{R}^2) \rtimes_\alpha \mathbf{R}$ has all fibers equal to $\mathcal{K}(L^2(\mathbf{R}))$, except at $x = 0$ where the fiber is $C_0(\mathbf{R}^2)$!

8.5.4

The continuous field in 8.5.3 is known as the **Heisenberg deformation**, the idea being that as $x \to 0$, the noncommutative algebra $\mathcal{K}(L^2(\mathbf{R}))$ is deformed continuously into the commutative algebra $C_0(\mathbf{R}^2)$. Thinking of x as Planck's constant, one might view this as the "passage to the classical limit" in quantum mechanics.

The name comes from a connection with the **Heisenberg group**, which is the subgroup H of $GL(3, \mathbf{R})$ consisting of all matrices of the form

$$\begin{pmatrix} 1 & x & y \\ 0 & 1 & t \\ 0 & 0 & 1 \end{pmatrix}$$

The subset N obtained by setting $t = 0$ is a normal subgroup isomorphic to $\mathbf{R}^2$. Setting $x = y = 0$ gives a subgroup K isomorphic to $\mathbf{R}$ that acts on N by conjugation:

$$\begin{pmatrix} 1 & 0 & 0 \\ 0 & 1 & t \\ 0 & 0 & 1 \end{pmatrix}^{-1} \begin{pmatrix} 1 & x & y \\ 0 & 1 & 0 \\ 0 & 0 & 1 \end{pmatrix} \begin{pmatrix} 1 & 0 & 0 \\ 0 & 1 & t \\ 0 & 0 & 1 \end{pmatrix} = \begin{pmatrix} 1 & x & tx + y \\ 0 & 1 & 0 \\ 0 & 0 & 1 \end{pmatrix}$$

So H can be described as a semidirect product $\mathbf{R}^2 \rtimes \mathbf{R}$ with the action of $\mathbf{R}$ on $\mathbf{R}^2$ being precisely that of the previous section. Thus 8.3.2 gives

$$C^*(H) \cong C^*(\mathbf{R}^2) \rtimes_\alpha \mathbf{R} \cong C_0(\mathbf{R}^2) \rtimes_\alpha \mathbf{R}$$

since $C^*(\mathbf{R}^2) \cong C_0(\hat{\mathbf{R}}^2)$ and $\hat{\mathbf{R}}^2 \cong \mathbf{R}^2$.

8.5.5

The **free product** $\mathbf{Z}_2 * \mathbf{Z}_2$ is, by definition, the group with generators a and b subject to the relations $a^2 = b^2 = 1$. It is amenable (the only free product of finite cyclic groups to be so), and it is possible to proceed as in 8.5.4 and give an explicit description of $C^*(\mathbf{Z}_2 * \mathbf{Z}_2)$. The subgroup N generated by ab is normal and isomorphic to $\mathbf{Z}$. The subgroup $H = \{1, a\} = \mathbf{Z}_2$ satisfies $H \cap N = \{1\}$ and together with N generates the full group. So $\mathbf{Z}_2 * \mathbf{Z}_2$ is a semidirect product $\mathbf{Z} \rtimes \mathbf{Z}_2$, and since $a^{-1}(ab)^n a = (ab)^{-n}$, the action of a on $\mathbf{Z}$ is its nontrivial automorphism $n \to -n$. Applying 8.3.2 gives

$$C^*(\mathbf{Z}_2 * \mathbf{Z}_2) \cong C^*(\mathbf{Z}) \rtimes_\alpha \mathbf{Z}_2 \cong C(S^1) \rtimes_\alpha \mathbf{Z}_2$$

where the generator of $\mathbf{Z}_2$ acts on S^1 by inversion (complex conjugation). Once again we have used the isomorphism $C^*(\mathbf{Z}) \cong C(\hat{\mathbf{Z}})$ mentioned in 3.6 and the fact that $\hat{\mathbf{Z}} \cong S^1$ (see 8.6.1).

Now the primitive ideal space of $C(S^1)$ is S^1 and the map

$$z \to \operatorname{Re} z : S^1 \to [-1, 1]$$

is continuous and open. So $C(S^1)$ is a continuous field over $[-1, 1]$ with fibers $\mathbf{C} \oplus \mathbf{C}$ over the points of the open interval and $\mathbf{C}$ over the endpoints. By 8.5.2, $C^*(\mathbf{Z}_2 * \mathbf{Z}_2)$ is a continous field over $[-1, 1]$ with fibers $(\mathbf{C} \oplus \mathbf{C}) \rtimes \mathbf{Z}_2 \cong M_2(\mathbf{C})$ over the open interval and $\mathbf{C} \rtimes \mathbf{Z}_2 \cong \mathbf{C}$ over the ends. It follows that

$$C^*(\mathbf{Z}_2 * \mathbf{Z}_2) \cong \{f \in C([-1, 1], M_2(\mathbf{C})) : f(-1), f(1) \text{ diagonal}\}$$

This subalgebra is generated by the self-adjoint unitaries

$$u = \begin{pmatrix} 1 & 0 \\ 0 & -1 \end{pmatrix} \quad \text{and} \quad v = \begin{pmatrix} \sin \frac{1}{2}\pi t & \cos \frac{1}{2}\pi t \\ \cos \frac{1}{2}\pi t & -\sin \frac{1}{2}\pi t \end{pmatrix}$$

which can be taken as the images of the generators a and b in $C^*(\mathbf{Z}_2 * \mathbf{Z}_2)$.

Alternatively, it is generated by the projections

$$p = \frac{1+u}{2} \quad \text{and} \quad q = \frac{1+v}{2}$$

It is in fact the universal C^*-algebra generated by two projections. This means that for any C^*-algebra A and projections $e, f \in A$, the assignment $p \to e$ and $q \to f$ extends to a $*$-homomorphism $C^*(\mathbf{Z}_2 * \mathbf{Z}_2) \to A$.

A third way to obtain this algebra is as the groupoid C^*-algebra associated to the equivalence relation on $[0, 1] \times \{1, 2\}$—two disjoint copies of $[0, 1]$—defined by identifying $(x, 1)$ and $(x, 2)$ for all $x \neq -1, 1$.

8.6 DUALITY

In this section we will consider the **duality theory** of locally compact groups and its connection with C^*-dynamical systems and crossed products. Duality theory works best for *abelian* groups, and we will restrict our considerations to this case (but note that a noncommutative theory does exist).

8.6.1

Let G be a locally compact Hausdorff abelian group, and consider the set $\hat{G}$ of **characters** of G, namely, the continuous homomorphisms from G to S^1, the circle group. With the pointwise product and the topology of uniform convergence on compact sets, $\hat{G}$ is again a locally compact abelian group, called the **dual group** or **character group** of G. A basic fact is that the canonical map from G to the dual of $\hat{G}$ (taking any $t \in G$ to evaluation at t) is an isomorphism of locally compact groups. Thus the duality relationship is symmetric.

A useful fact is that compact groups correspond via duality to discrete groups. For example, any continuous homomorphism $S^1 \to S^1$ is of the form $e^{i\theta} \to e^{in\theta}$ for some $n \in \mathbf{Z}$; hence $\hat{S}^1 \cong \mathbf{Z}$. It follows that $\hat{\mathbf{Z}} \cong S^1$ (or it can be proved directly that homomorphisms $\mathbf{Z} \to S^1$ are of the form $n \to e^{in\theta}$ for some $e^{i\theta} \in S^1$). This was used in 8.5.5.

At several points in 8.5 we made use of the fact that the group $\mathbf{R}$ of real numbers with the usual topology is self-dual: $\hat{\mathbf{R}} \cong \mathbf{R}$. This amounts to the assertion that any continuous homomorphism $\mathbf{R} \to S^1$ has the form $r \to e^{irt}$ for some $t \in \mathbf{R}$. It can also be noted that any finite abelian group is self-dual.

8.6.2

In 3.6 we mentioned and in 8.5 we made use of the fact that $C^*(G) \cong C_0(\hat{G})$ for any locally compact abelian group G. To see this, the first thing to observe is that if G is abelian, then the convolution product is commutative: For all $f, g \in C_c(G)$ we have

$$\int_G f(s)g(s^{-1}t)ds = \int_G f(st)g(s^{-1})ds = \int_G f(s^{-1}t)g(s)ds$$

using the fact that abelian groups are unimodular. Hence $C^*(G)$ is a commutative C^*-algebra, and making use of the Gelfand-Naimark Theorem (2.2.4), we need to identify $\hat{G}$ with the space of nonzero multiplicative linear functionals on $C^*(G)$ in the relative weak*-topology. Given a multiplicative functional on $C^*(G)$, regard it as a one-dimensional representation, extend it to a representation of the multiplier algebra of $C^*(G)$ by 5.8.1, and restrict this to the canonical copy of G. The result is a one-dimensional unitary representation of G, or in other words an element of $\hat{G}$. To see that this is a bijection, we can make use of the formula in 8.3.2: For a character τ of G,

$$\phi(f) = \hat{f}(\tau) = \int_G f(s)\tau(s)ds, \quad f \in C_c(G)$$

defines a nonzero multiplicative linear functional ϕ on $C_c(G)$, hence on $C^*(G)$.

8.6.3

Let (G, α, A) be a C^*-dynamical system with G abelian. To any $\tau \in \hat{G}$ we associate the map

$$f \to \tau f : C_c(G, A) \to C_c(G, A)$$

where τf is just the pointwise product

$$(\tau f)(t) = \tau(t) f(t) \qquad \text{for all } t \in G$$

In fact this is a $*$-homomorphism:

$$\begin{aligned}((\tau f) * (\tau g))(t) &= \int_G \tau(s) f(s)\alpha_s(\tau(s^{-1}t)g(s^{-1}t))ds \\ &= \tau(t) \int_G f(s)\alpha_s(g(s^{-1}t))ds \\ &= (\tau(f * g))(t)\end{aligned}$$

$$(\tau f)^*(t) = \alpha_t((\tau f)(t^{-1})^*) = \alpha_t(\overline{\tau(t^{-1})} f(t^{-1})^*) = (\tau f^*)(t)$$

It is evidently bounded and so extends to $A \rtimes_\alpha G$ to yield a map

$$\hat{\alpha} : \hat{G} \to \mathrm{Aut}(A \rtimes_\alpha G)$$

which is clearly a homomorphism. We therefore have a dynamical system $(\hat{G}, \hat{\alpha}, A \rtimes_\alpha G)$, called the **dual C^*-dynamical system** of the given system (G, α, A).

8.6.4

As an example, consider the dual action $\hat{\alpha}$ of $\hat{G}$ on $C^*(G)$. Recalling (8.6.2) that $C^*(G) \cong C_0(\hat{G})$, what does this action look like on $C_0(\hat{G})$? As indicated there, this isomorphism takes $f \in C_c(G)$ to $\hat{f} \in C_0(\hat{G})$ defined by $\hat{f}(\tau) = \int_G f(s)\tau(s)ds$. From this it is clear that the action of $\hat{G}$ on $C_0(\hat{G})$ is induced by "right translation," the action $(\sigma, \tau) \to \sigma\tau$ of $\hat{G}$ on itself. This is reminiscent of the situation in 8.4.3 where, with the action $(s, t) \to s^{-1}t$ of G on itself, we established that $C_0(G) \rtimes_\gamma G \cong \mathcal{K}(L^2(G))$. Indeed there are isomorphisms

$$C^*(G) \rtimes_{\hat{\alpha}} \hat{G} \cong C^*(\hat{G}) \rtimes_\beta G \cong C_0(G) \rtimes_\gamma G$$

where $(\beta_t f)(\tau) = \overline{\tau(t)} f(\tau)$ for all $f \in C_c(\hat{G})$, $\tau \in \hat{G}$, and $t \in G$. In other words,

$$(\mathbf{C} \rtimes_\alpha G) \rtimes_{\hat{\alpha}} \hat{G} \cong \mathcal{K}(L^2(G))$$

where α is the trivial action. This is a special case of a general result

$$(A \rtimes_\alpha G) \rtimes_{\hat{\alpha}} \hat{G} \cong A \otimes \mathcal{K}(L^2(G))$$

to be discussed in the next section. Here we recall that $\mathcal{K}$ is nuclear and that the tensor product denotes the completion in the unique C^*-norm.

The isomorphisms mentioned in the previous paragraph are easy to establish. For the first of these the algebras in question are the norm closures of $C_c(\hat{G} \times G)$ and $C_c(G \times \hat{G})$, regarded as subalgebras of $C_c(\hat{G}, C_c(G))$ and $C_c(G, C_c(\hat{G}))$ and as acting on $L^2(\hat{G} \times G)$ and $L^2(G \times \hat{G})$, respectively. Applying 8.2.4 with (G, α, A) replaced by $(\hat{G}, \hat{\alpha}, C^*(G))$ and with π the regular representation of $C^*(G)$ on $L^2(G)$, we find (identifying $f \in C_c(\hat{G} \times G)$ with the corresponding multiplication operator on $L^2(\hat{G} \times G)$) that

$$(f\xi)(\tau, t) = \int_{\hat{G}} \int_G \overline{\tau(s)} f(\sigma, s)\xi(\sigma^{-1}\tau, s^{-1}t)ds\,d\sigma$$

for all $\xi \in L^2(\hat{G} \times G)$. Likewise applying 8.2.4 to $(G, \beta, C^*(\hat{G}))$ with π the regular representation of $C^*(\hat{G})$ on $L^2(\hat{G})$ gives

$$(g\eta)(t, \tau) = \int_G \int_{\hat{G}} \sigma(t) g(s, \sigma)\eta(s^{-1}t, \sigma^{-1}\tau)d\sigma\,ds$$

for all $g \in C_c(G \times \hat{G})$ and $\eta \in L^2(G \times \hat{G})$. Next we define a unitary operator W from $L^2(G \times \hat{G})$ to $L^2(\hat{G} \times G)$ by

$$(W\eta)(\tau, t) = \overline{\tau(t)}\eta(t, \tau)$$

and compute that

$$(W^* f W\eta)(t, \tau) = \int_{\hat{G}} \int_G \sigma(t)[\overline{\sigma(s)} f(\sigma, s)]\eta(s^{-1}t, \sigma^{-1}\tau)dsd\sigma$$

for all $f \in C_c(\hat{G} \times G)$ and $\eta \in L^2(G \times \hat{G})$. In other words, if we define $g \in C_c(G \times \hat{G})$ by

$$g(t, \tau) = \overline{\tau(t)} f(\tau, t)$$

then $W^* f W = g$. Since g runs over $C_c(G \times \hat{G})$ as f runs over $C_c(\hat{G} \times G)$, we conclude that

$$C^*(G) \rtimes_{\hat{\alpha}} \hat{G} \cong C^*(\hat{G}) \rtimes_{\beta} G$$

as claimed.

8.6.5

The second isomorphism mentioned in 8.6.4 is a consequence of the fact that the C^*-dynamical systems

$$(G, \beta, C^*(\hat{G})) \quad \text{and} \quad (G, \gamma, C_0(G))$$

are covariantly isomorphic, in the sense of 8.3.3. In fact the isomorphism of 8.6.2 (with G replaced by $\hat{G}$) takes $f \in C_c(\hat{G})$ to $\hat{f} \in C_0(G)$ defined by

$$\hat{f}(t) = \int_{\hat{G}} f(\sigma)\sigma(t)d\sigma$$

and we compute

$$\widehat{\beta_s f}(t) = \int_{\hat{G}} \overline{\sigma(s)} f(\sigma)\sigma(t)d\sigma = \hat{f}(s^{-1}t) = (\gamma_s \hat{f})(t)$$

as asserted.

8.6.6

Given (G, α, A) with G abelian, just as there is an action $\hat{\alpha}$ of $\hat{G}$ on $A \rtimes_\alpha G$, and hence a dynamical system $(\hat{G}, \hat{\alpha}, A \rtimes_\alpha G)$, there is an action $\hat{\hat{\alpha}}$ of $\hat{\hat{G}}$ on $A \rtimes_\alpha G \rtimes_{\hat{\alpha}} \hat{G}$. Since $\hat{\hat{G}} = G$ it is natural to ask about the relationship of $A \rtimes_\alpha G \rtimes_{\hat{\alpha}} \hat{G}$ to A (and of $\hat{\hat{\alpha}}$ to α). We have already shown that

$$\mathbf{C} \rtimes_\alpha G \rtimes_{\hat{\alpha}} \hat{G} = C^*(G) \rtimes_{\hat{\alpha}} G \cong \mathcal{K}(L^2(G))$$

The general result is about as simple as it could be, compatible with this. It is due to Takai [Tak75]. To state it, we will need the **right regular representation** ρ of G on $L^2(G)$:

$$(\rho_s \xi)(t) = \xi(ts) \qquad \text{for all } \xi \in L^2(G) \text{ and } s, t \in G$$

This is a unitary representation since G is abelian and hence unimodular.

8.6.7 ***Theorem*** (Takai duality)

For any C^*-dynamical system (G, α, A) with G abelian, there is a covariant isomorphism

$$(G, \hat{\hat{\alpha}}, (A \rtimes_\alpha G) \rtimes_{\hat{\alpha}} \hat{G}) \cong (G, \alpha \otimes \operatorname{Ad} \rho, A \otimes \mathcal{K}(L^2(G)))$$

of the double dual system, where ρ is the right regular representation of G.

Outline of the proof

The proof (see [Ped79, 7.9.3]) relies heavily on the ideas of the preceding sections. The main new idea is that

$$(G, \alpha \otimes \operatorname{Ad} \rho, A \otimes \mathcal{K}(L^2(G))) \quad \text{and} \quad (G, \delta, C_0(G, A) \rtimes_\gamma G)$$

are covariantly isomorphic, where γ is given by (generalizing 8.4.3)

$$(\gamma_s f)(t) = \alpha_s(f(s^{-1}t)) \qquad \text{for } f \in C_0(G, A) \text{ and } s, t \in G$$

and δ by

$$(\delta_s f)(t, r) = f(t, rs) \qquad \text{for } f \in C_c(G \times G, A) \text{ and } r, s, t \in G$$

The proof is then completed by arguing, just as in 8.6.4 and 8.6.5, that

$$C_0(G, A) \rtimes_\gamma G \cong (A \otimes C^*(\hat{G})) \rtimes_\beta G \cong (A \rtimes_\alpha G) \rtimes_{\hat{\alpha}} \hat{G}$$

and that in addition these isomorphisms carry the action δ of G to $\hat{\hat{\alpha}}$.

Start with a faithful representation of A on a Hilbert space $\mathcal{H}$. We can assume that $A \subset \mathcal{B}(\mathcal{H})$ and hence that $C_0(G, A) \subset \mathcal{B}(L^2(G, H))$. As in 8.2.4 we then have faithful representation of $C_0(G, A) \rtimes_\gamma G$ on $L^2(G \times G, \mathcal{H})$ given by

$$(f\xi)(t, \cdot) = \int_G \gamma_{t^{-1}}(f(s, \cdot))\xi(s^{-1}t, \cdot)ds$$

for all $f \in C_c(G \times G, A)$ and $\xi \in L^2(G \times G, \mathcal{H})$, where we have identified f with $(\tilde{\pi} \times \tilde{\lambda})f$. Hence

$$(f\xi)(t, r) = \int_G \alpha_{t^{-1}}(f(s, tr)\xi(s^{-1}t, r)ds$$

from the definition of γ.

On the other hand, if we represent A faithfully on $L^2(G, \mathcal{H})$ by

$$(a\xi)(t) = \alpha_t(a)\xi(t) \qquad \text{for all } \xi \in L^2(G, A)$$

then we may regard $A \otimes \mathcal{K}(L^2(G))$ as acting on

$$L^2(G, \mathcal{H}) \otimes L^2(G) \cong L^2(G \times G, \mathcal{H})$$

as well. Define a unitary operator W on $L^2(G \times G, \mathcal{H})$ by

$$(W\xi)(t, r) = \xi(tr, r) \qquad \text{for all } \xi \in L^2(G \times G, \mathcal{H})$$

and compute, for all $f \in C_c(G \times G, A)$,

$$\begin{aligned}(W^* f W\xi)(t, r) &= (f W\xi)(tr^{-1}, r) \\ &= \int_G \alpha_{rt^{-1}}(f(s, t))W\xi(s^{-1}tr^{-1}, r)ds \\ &= \int_G \alpha_{rt^{-1}}(f(s, t))\xi(s^{-1}t, r)ds\end{aligned}$$

Proceeding as in 8.4.3, let $g, h \in C_c(G)$ and $a \in A$, and take

$$f(s, t) = g(s^{-1}t)h(t)\alpha_t(a)$$

in this formula to get

$$\begin{aligned}(W^* f W\xi)(t, r) &= \int_G g(s^{-1}t)h(t)\alpha_r(a)\xi(s^{-1}t, r)ds \\ &= h(t)\alpha_r(a) \int_G g(s)\xi(s, r)ds\end{aligned}$$

Here we let $\xi = \eta \otimes \zeta \in L^2(G, \mathcal{H}) \otimes L^2(G)$, and so

$$W^* f W(\eta \otimes \zeta) = a\eta \otimes \langle \zeta, \overline{g} \rangle h$$

It follows from this that W implements a spatial isomorphism between $C_0(G, A) \rtimes_\gamma G$ and $A \otimes \mathcal{K}(L^2(G))$.

To see that this is an isomorphism of dynamical systems, replace f by $\delta_u f$, $u \in G$, and repeat the computation to get

$$\begin{aligned}(W^*(\delta_u f)W\xi)(t, r) &= h(tu)\alpha_{tu}(a) \int_G g(su)\xi(s, r)ds \\ &= (\rho_u h)(t)\alpha_t(\alpha_u(a)) \int_G (\rho_u g)(s)\xi(s, r)ds,\end{aligned}$$

so that if $\xi = \eta \otimes \zeta$ as before, then

$$\begin{aligned}W^*(\delta_u f)W(\eta \otimes \zeta) &= \alpha_u(a)\eta \otimes \langle \zeta, \overline{\rho_u g} \rangle \rho_u h \\ &= \alpha_u(a)\eta \otimes \langle \rho_u^* \zeta, \overline{g} \rangle \rho_u h \\ &= (\alpha_u \otimes \mathrm{Ad}\, \rho_u)(a \otimes \langle \cdot, \overline{g} \rangle h)(\eta \otimes \zeta)\end{aligned}$$

Hence $W^*(\delta_u f)W = (\alpha_u \otimes \mathrm{Ad}\rho_u) f$ as required.

Raeburn has given an interesting non-spatial proof of this result, using the universal property of the crossed product [Rae88].

8.6.8

C^*-algebras A and B are said to be **stably isomorphic** if the C^*-algebras $A \otimes \mathcal{K}$ and $B \otimes \mathcal{K}$ are $*$-isomorphic, where $\mathcal{K}$ is the algebra of compact operators on a separable infinite dimensional Hilbert space (and the tensor product denotes the completion in the unique C^*-norm). The relation of stable isomorphism is frequently relevant, as for example in Theorem 8.6.7 which says that, up to stable isomorphism, the C^*-dynamical systems (G, α, A) and $(\hat{G}, \hat{\alpha}, A \rtimes_\alpha G)$ are duals of each other. We will see in Chapter 10 that it arises naturally in the classification theory of C^*-algebras via K-theory.

Isomorphic C^*-algebras are stably isomorphic but the converse is false. This is clear because algebras of the form $A \otimes \mathcal{K}$ are never unital, but here is a more informative example: Since $\mathcal{K}(\mathcal{H}) \otimes \mathcal{K}(\mathcal{H})$ can be identified as the compact operators on $\mathcal{H} \otimes \mathcal{H}$, we see that $\mathcal{K} \otimes \mathcal{K} \cong \mathcal{K}$ and hence that $\mathbf{C}$ and $\mathcal{K}$ are stably isomorphic. A C^*-algebra A is said to be **stable** if $A \otimes \mathcal{K} \cong A$; for such algebras isomorphism and stable isomorphism coincide.

8.7 W*-DYNAMICAL SYSTEMS

The concept of a **W^*-dynamical system** $(G, \alpha, \mathcal{M})$ was considered briefly in 4.9 in the course of describing the Murray and von Neumann construction of factors of types II and III. There $\mathcal{M}$ was taken to be abelian and G discrete, but in general $\mathcal{M}$ can be any von Neumann algebra and G any locally compact topological group. In addition we require of the homomorphism α from G to Aut $\mathcal{M}$ that $t \to \alpha_t(x)$ be σ-weakly continuous (see 4.4). If we assume that $\mathcal{M}$ is a von Neumann algebra on $\mathcal{H}$, then the **W^*-crossed product** $\mathcal{M} \rtimes_\alpha G$ can be described as the von Neumann algebra on $L^2(G, \mathcal{H})$ generated by the operators

$$(a\xi)(t) = \alpha_{t^{-1}}(a)\xi(t) \quad \text{and} \quad (\tilde{\lambda}_s\xi)(t) = \xi(s^{-1}t)$$

of 8.2.4 for all $a \in \mathcal{M}$ and $s \in G$, where $\xi \in L^2(G, \mathcal{H})$. As mentioned there in the C^*-algebra case, a straightforward computation shows that these equations define a covariant representation of the W*-dynamical system. In particular, $\mathcal{M} \rtimes_\alpha G$ contains canonical copies of $\mathcal{M}$ and G that generate it as a von Neumann algebra. It is also generated by the operators

$$(f\xi)(t) = \int_G \alpha_{t^{-1}}(f(s))\xi(s^{-1}t)ds$$

for all $f \in C_c(G, \mathcal{M})$.

8.7.1

Given a W^*-dynamical system $(G, \alpha, \mathcal{M})$ with $\mathcal{M}$ acting on $\mathcal{H}$, the **dual W^*-dynamical system** $(\hat{G}, \hat{\alpha}, \mathcal{M} \rtimes_\alpha G)$ is defined, as in the case of C^*-dynamical systems, by letting $\hat{G}$ act on $C_c(G, \mathcal{M})$ by pointwise multiplication. For $\tau \in \hat{G}$ and $f \in C_c(G, \mathcal{M})$, the action of $\hat{\alpha}_\tau(f) = \tau f$ on $L^2(G, \mathcal{H})$ is given by

$$(\hat{\alpha}_\tau(f)\xi)(t) = \int_G \tau(s)\alpha_{t^{-1}}(f(s))\xi(s^{-1}t)ds$$

From this it follows that $\hat{\alpha}_\tau(f) = (\operatorname{Ad} u_\tau)f$, where $\tau \to u_\tau$ is the unitary representation of $\hat{G}$ on $L^2(G, \mathcal{H})$ defined by $u_\tau\xi = \tau\xi$ (pointwise product), and hence that $\hat{\alpha}_\tau$ extends to all of $\mathcal{M} \rtimes_\alpha G$.

8.7.2 ***Theorem*** (Takesaki duality)

For any W^*-dynamical system $(G, \alpha, \mathcal{M})$ with $\mathcal{M}$ abelian, there is a covariant isomorphism

$$(G, \hat{\hat{\alpha}}, (\mathcal{M} \rtimes_\alpha G) \rtimes_\alpha \hat{G}) \cong (G, \alpha \otimes \operatorname{Ad} \rho, \mathcal{M} \otimes \mathcal{B}(L^2(G)))$$

This result is identical to the C^*-algebra version, except that $\mathcal{K}(L^2(G))$ has been replaced by its weak operator closure $\mathcal{B}(L^2(G))$ and the tensor product is the von Neumann algebra tensor product (see 4.3.10). It can be deduced from the C^*-algebra version (see [Ped79, 7.10.6]), but in fact it was discovered first, by Takesaki [Tak73].

8.7.3

Consider a W^*-crossed product $\mathcal{N} = \mathcal{M} \rtimes_\alpha G$ with G abelian and $\mathcal{M}$ a von Neumann algebra on $\mathcal{H}$ so that $\mathcal{N} \subset \mathcal{B}(L^2(G, \mathcal{H}))$. From the foregoing we have representations $\tilde{\lambda}$ of G in $\mathcal{U}(\mathcal{N})$ and $\hat{\alpha}$ of $\hat{G}$ in $\operatorname{Aut} \mathcal{N}$. Moreover, recalling that $\hat{\alpha}_\tau = \operatorname{Ad} u_\tau$, we have

$$\begin{aligned}
(\hat{\alpha}_\tau(\tilde{\lambda}_s)\xi)(t) &= (u_\tau\tilde{\lambda}_s u_{\tau^{-1}}\xi)(t) \\
&= \tau(t)(u_\tau^{-1}\xi)(s^{-1}t) \\
&= \tau(t)\tau^{-1}(s^{-1}t)\xi(s^{-1}t) \\
&= \tau(s)(\tilde{\lambda}_s\xi)(t)
\end{aligned}$$

so that $\hat{\alpha}_\tau(\tilde{\lambda}_s) = \tau(s)\tilde{\lambda}_s$, and

$$\begin{aligned}(\hat{\alpha}_\tau(a)\xi)(t) &= (u_\tau a u_{\tau^{-1}}\xi)(t) \\ &= \tau(t)\alpha_{\tau^{-1}}(a)(u_{\tau^{-1}}\xi)(t) \\ &= \alpha_{t^{-1}}(a)\xi(t) \\ &= (a\xi)(t)\end{aligned}$$

Thus $\hat{\alpha}_\tau$ leaves each element of $\mathcal{M}$ fixed.

These observations lead to a characterization of W^*-crossed products by abelian groups: If $\mathcal{N}$ is a von Neumann algebra and G is a locally compact abelian group for which there are representations $\tilde{\lambda}$ of G in $\mathcal{U}(\mathcal{N})$ and $\hat{\alpha}$ of $\hat{G}$ in Aut $\mathcal{N}$ such that $\hat{\alpha}_\tau(\tilde{\lambda}_s) = \tau(s)\tilde{\lambda}_s$ for all $\tau \in \hat{G}$ and $s \in G$, then $\mathcal{N}$ is a W^*-crossed product $\mathcal{M} \rtimes_\alpha G$, where $\mathcal{M}$ is the fixed point algebra of $\hat{\alpha}$ and $\alpha = \operatorname{Ad}\lambda$ [Ped79, 7.10.4]. Moreover the system $(G, \alpha, \mathcal{M})$ is unique up to covariant isomorphism.

Chapter Nine

Factors

9.1 THE STANDARD REPRESENTATION

Factors were introduced in 4.6.3: A **factor** is a von Neumann algebra $\mathcal{M}$ that cannot be decomposed as a direct sum of (nonzero) von Neumann algebras. Murray and von Neumann showed that any von Neumann algebra can be decomposed, in a suitable sense, into a "direct integral" of factors, thereby reducing (in principle) questions about general algebras to the case of factors. They described the type classification of factors (4.7), showed that type I factors are spatially isomorphic to an inflation $\mathcal{B}(\mathcal{H})^{(n)}$ (and hence isomorphic to $\mathcal{B}(\mathcal{H})$) and gave examples of factors of type II and type III (4.9.1; 4.9.4). They introduced the important class of hyperfinite algebras and demonstrated the uniqueness of the hyperfinite type II_1 factor (4.10). The creation in 1967–70 of the Tomita-Takesaki modular automorphism theory heralded a new stage in the development of von Neumann algebra theory. Subsequent work, led by Alain Connes and Vaughan Jones but including significant contributions by many others, has produced a revolution in the study of factors. We describe some of these developments in this chapter.

9.1.1

Let $\mathcal{M}$ be a finite factor. Then the unique normalized trace on $\mathcal{M}$ is a state and applying the GNS construction yields the **standard representation** of $\mathcal{M}$. We will denote the trace by tr, the representation space $\mathcal{H}_{\text{tr}}$ by $L^2(\mathcal{M})$ and the associated cyclic vector by ξ_0. This representation is faithful (because the trace is), and its range is weakly closed (because the trace is σ-weakly continuous); therefore we can identify $\mathcal{M}$ with its image in $\mathcal{B}(L^2(\mathcal{M}))$. This realization of $\mathcal{M}$, called the **standard form** of $\mathcal{M}$, has many useful properties, mainly resulting from the fact that the cyclic vector ξ_0 is also separating and tracial:

i. $a \in \mathcal{M}$ and $a\xi_0 = 0 \Rightarrow a = 0$.
ii. $\langle ab\xi_0, \xi_0\rangle = \langle ba\xi_0, \xi_0\rangle$ for all $a, b \in \mathcal{M}$.

These properties follow trivially from the equation 5.3.1(i) relating a state and its GNS representation, which here reads $\mathrm{tr}(a) = \langle a\xi_0, \xi_0\rangle$.

9.1.2 *Lemma*

Let $\mathcal{M} \subset \mathcal{B}(\mathcal{H})$ be a von Neumann algebra having a cyclic and separating trace vector $\xi_0 \in \mathcal{H}$. Then the map $J : \mathcal{M}\xi_0 \to \mathcal{M}\xi_0$ defined by

$$J(a\xi_0) = a^*\xi_0 \qquad \text{for all } a \in \mathcal{M}$$

extends uniquely to a conjugate linear isometry of $\mathcal{H}$ onto itself. Moreover $J^2 = I$, the identity operator.

Proof

Since ξ_0 is separating the definition of J is unambiguous. Because ξ_0 is tracial, J is an isometry, so by cyclicity it extends by continuity to an isometry of $\mathcal{H}$ onto itself.

9.1.3 *Proposition*

With the notation of 9.1.2, let θ be the $*$-automorphism of $\mathcal{B}(\mathcal{H})$ given by

$$\theta(a) = JaJ \qquad \text{for all } a \in \mathcal{B}(\mathcal{H})$$

Then $\theta(\mathcal{M}) = \mathcal{M}'$, the commutant of $\mathcal{M}$.

Proof

For all $a, b \in \mathcal{M}$ we have

$$\theta(a)b\xi_0 = Jab^*\xi_0 = ba^*\xi_0$$

Applying this twice gives

$$a_1\theta(a_2)b\xi_0 = a_1ba_2^*\xi_0 = \theta(a_2)a_1b\xi_0$$

and hence $\theta(\mathcal{M}) \subset \mathcal{M}'$. Equality holds here because this situation is completely symmetric with respect to $\mathcal{M}$ and $\mathcal{M}'$. It is an elementary fact that a vector is cyclic for a von Neumann algebra if and only if it is separating for the commutant. Thus ξ_0 is cyclic and separating for $\mathcal{M}'$, and so the map $a'\xi_0 \to a'^*\xi_0 (a' \in \mathcal{M}')$ is unambiguously and densely defined. In fact it is just the map J again. To see this, fix $a' \in \mathcal{M}'$, and choose $\{a_n\} \subset \mathcal{M}$ with $a_n\xi_0 \to a'\xi_0$. If we show that $a_n^*\xi_0 \to a'^*\xi_0$, it will follow that

$$J(a'\xi_0) = \lim J(a_n\xi_0) = \lim a_n^*\xi_0 = a'^*\xi_0$$

as claimed. Now $a_n^*\xi_0 \to a'^*\xi_0$ weakly:

$$\langle a_n^*\xi_0, b\xi_0\rangle = \langle \xi_0, a_n b\xi_0\rangle = \langle \xi_0, ba_n\xi_0\rangle \to \langle \xi_0, ba'\xi_0\rangle = \langle a'^*\xi_0, b\xi_0\rangle$$

for all $b \in \mathcal{M}$, $\mathcal{M}\xi_0$ is dense and $\{a_n^*\xi_0\}$ is bounded (since J is isometric). But $\{a_n^*\xi_0\}$ is a norm Cauchy sequence, again since J is isometric, so $a_n^*\xi_0 \to a'\xi_0$ in norm. We can now repeat the argument at the beginning of the proof to see that $\theta(\mathcal{M}') \subset \mathcal{M}$. But θ^2 is the identity map, and the result follows.

9.2 THE TOMITA-TAKESAKI MODULAR THEORY

In 1967 Tomita considered an extension of these ideas to arbitrary algebras. As clarified by Takesaki [Tak70], the resulting **Tomita-Takesaki modular theory** has had far-reaching consequences.

9.2.1

To include more general algebras, especially those of type III, the idea is to replace the possibly nonexistent traces by faithful normal states (or even weights). Any von Neumann algebra $\mathcal{M}$ on a separable Hilbert space has such a state. For example, take $\varphi(a) = \sum 2^{-n}\langle a\xi_n, \xi_n\rangle$, where $\{\xi_n\}$ is an orthonormal basis. The GNS representation will be faithful if the state is, and its range will be a von Neumann algebra if the state is normal. So as in 9.1.1 we can identify a von Neumann algebra $\mathcal{M}$ with its image in the GNS representation corresponding to a faithful normal state φ; equivalently, we can suppose that $\mathcal{M}$ has a cyclic and separating vector ξ_0.

This time, however, because ξ_0 is not necessarily a trace vector, the map $a\xi_0 \to a^*\xi_0$ $(a \in \mathcal{M})$ need not be an isometry, and in fact it need not even be bounded. However, Tomita was able to show that it is always closable. This is significant because the polar decomposition is available for any closed densely defined operator. Taking the polar decomposition $J\Delta^{1/2}$ of the closure we again get a conjugate linear isometry J with the same properties as before.

The closed self-adjoint operator Δ is called the **modular operator**. It incorporates further important information about the algebra $\mathcal{M}$. A fundamental fact is that the automorphisms $x \to \Delta^{it} x \Delta^{-it}$, $t \in \mathbf{R}$, of $\mathcal{B}(\mathcal{H})$ restrict to automorphisms of $\mathcal{M}$. In other words, from φ one gets an action $\sigma^{\varphi} : \mathbf{R} \to \operatorname{Aut} \mathcal{M}$ and hence a W*-dynamical system $(\mathbf{R}, \sigma^{\varphi}, \mathcal{M})$. This one-parameter group of automorphisms of $\mathcal{M}$ is called the **modular automorphism group** (of φ).

As indicated, this process admits a further generalization, from states on $\mathcal{M}$ to weights. To ensure that the GNS representation corresponding to a weight is faithful, it is necessary to assume that the weight is semifinite as well as faithful. Thus to any faithful normal semifinite weight φ on a von Neumann algebra is associated a conjugation J_{φ}, a modular operator Δ_{φ}, and a modular automorphism group σ^{φ} with the same properties as before.

What other properties does the modular automorphism group have? It can be shown (see [KaR86, 9.2] or, for a sketch, [Tak83]) that it is the unique one-parameter automorphism group σ_t with the following properties:

i. $\varphi \circ \sigma_t = \varphi$ for all $t \in \mathbf{R}$.
ii. If $a, b \in \{x : \varphi(x^*x), \varphi(xx^*) < \infty\}$, there is a complex-valued function F, continuous on the horizontal strip $\{z : 0 \leq \operatorname{Im} z \leq 1\}$ and holomorphic in the interior, such that $F(t) = \varphi(\sigma_t(a)b)$ and $F(t+i) = \varphi(b\sigma_t(a))$ for all $t \in \mathbf{R}$.

The second property is the **KMS** (Kubo-Martin-Schwinger) **boundary condition** from statistical mechanics. As an example of how this result can be used, it is easy to see that if $h \in \mathcal{B}(\mathcal{H})$ is positive and invertible then the weight $\varphi(x) = \operatorname{tr}(hx)$ and the automorphism group $\sigma_t = \operatorname{Ad}(h^{it})$ satisfy conditions i and ii on $\mathcal{M} = \mathcal{B}(\mathcal{H})$; hence $\sigma^{\varphi} = \sigma$.

9.2.2

Connes has investigated the dependence of the modular automorphism group on the weight. To approach this, let $\{\sigma_t\}$ be a one-parameter group of automorphisms of a von Neumann algebra $\mathcal{M}$, let $\{u_t\}$ be a one-parameter family of unitary elements of $\mathcal{M}$, and consider whether the composition $(\operatorname{Ad} u) \circ \sigma$ is a one-parameter group. A simple computation shows that a natural condition guaranteeing this is the **cocycle condition**

$$u_{s+t} = u_s \sigma_s(u_t) \qquad \text{for all } s, t \in \mathbf{R}$$

Then we have [Con73]:

Theorem

If φ and ψ are faithful normal semifinite weights on a von Neumann algebra $\mathcal{M}$, then there is a σ-weakly continuous one-parameter family $\{u_t\}$ of unitaries in $\mathcal{M}$ satisfying the cocycle condition with respect to φ such that $\sigma^\psi = (\mathrm{Ad}\, u) \circ \sigma^\varphi$.

The proof of this uses the elegant "2×2 matrix trick". If χ is the weight on $M_2(\mathcal{M})$ defined by $\chi(x) = \varphi(x_{11}) + \psi(x_{22})$, then it is not difficult to see that $\sigma_t^\chi(e_{12})$ has the form $\left(\begin{smallmatrix} 0 & u_t \\ 0 & 0 \end{smallmatrix}\right)$ for some one-parameter family $\{u_t\}$ of unitaries in $\mathcal{M}$, where e_{12} is the $(1, 2)$ matrix unit in $M_2(\mathcal{M})$, and that this family has the required properties (see [Tak83]).

Thus the modular automorphism group is essentially (up to inner automorphisms) independent of the weight, and we see that every von Neumann algebra has associated with it a canonical flow (action of $\mathbf{R}$).

Associated with the modular automorphism group are the following two isomorphism invariants of the algebra:

$$S(\mathcal{M}) = \bigcap \{\sigma(\Delta_\varphi) : \varphi \text{ a faithful normal semifinite weight on } \mathcal{M}\}$$

$$T(\mathcal{M}) = \{t \in \mathbf{R} : \sigma_t \text{ is inner}\}$$

By the theorem one can take in the definition of T the modular automorphism group associated to any faithful normal semifinite weight. This invariant is significant only for type III algebras: $\mathcal{M}$ is semifinite if and only if $T(\mathcal{M}) = \mathbf{R}$ [Tak70]. The other invariant was introduced by Connes [Con73], who showed that it is a closed multiplicative semigroup of $\mathbf{R}_+$ so that there are only the following possibilities:

$$S(\mathcal{M}) = \begin{cases} [0, +\infty) & \\ \{\lambda^n : n \in \mathbf{Z}\} \cup \{0\} & \text{for some } \lambda \in (0, 1) \\ \{0, 1\} & \end{cases}$$

We say that $\mathcal{M}$ is, respectively, **type III_0**, **type III_λ**, or **type III_1**. For example, the Powers factors $\mathcal{R}_t$ (5.3.6) are of type III_λ, where $\lambda = t/(1 - t)$ (for t strictly between 0 and $\frac{1}{2}$), and so they are pairwise nonisomorphic.

Just as in the type II case (see 4.10), there are strong uniqueness results for hyperfinite (i.e., amenable) type III factors. Thus for each $\lambda \in (0, 1)$ there is, up to isomorphism, exactly one hyperfinite III_λ factor, namely the Powers factor. The hyperfinite III_1 factor, which occurs in a class of factors constructed by

H. Araki and J. Woods using an extension of Powers's method, is also unique. There is no uniqueness in the III_0 case, and a still more general construction, provided by W. Kreiger, is needed to obtain all of them. The result is that any hyperfinite III_0 factor is isomorphic to a Kreiger factor; the Kreiger factors are distinguished by an additional invariant, the "flow of weights." For details and references the reader should consult Connes's book [Con94, ch. 5].

9.2.3

The following question is of interest: If $\mathcal{M}$ is a II_1 factor and p is a projection in $\mathcal{M}$, is $p\mathcal{M}p \cong \mathcal{M}$? It is related to the existence of automorphisms of the II_∞ factor $\mathcal{M} \otimes \mathcal{B}(\mathcal{H})$ that "scale the trace." The canonical trace in $\mathcal{M}$ extends in an obvious way to a faithful semifinite trace τ (4.8.1) on $\mathcal{M} \otimes \mathcal{B}(\mathcal{H})$, and the uniqueness of the normalized trace in $\mathcal{M}$ implies that faithful semifinite traces in $\mathcal{M} \otimes \mathcal{B}(\mathcal{H})$ agree up to a scale factor. Thus any automorphism α of $\mathcal{M} \otimes \mathcal{B}(\mathcal{H})$ satisfies $\tau \circ \alpha = \lambda\tau$ for some $\lambda \in (0, +\infty)$. Suppose that $\lambda < 1$. Then a projection p as above can be constructed as follows: fixing a rank 1 projection e_0 in $\mathcal{B}(\mathcal{H})$, we have $\tau(\alpha(I \otimes e_0)) < \tau(I \otimes e_0)$, so

$$\alpha(I \otimes e_0) \sim p \otimes e_0 \qquad \text{for some } p \in \mathcal{M},\ \tau(p) = \lambda$$

and therefore

$$\begin{aligned} \mathcal{M} &\cong (I \otimes e_0)(\mathcal{M} \otimes \mathcal{B}(\mathcal{H}))(I \otimes e_0) \\ &\cong (p \otimes e_0)(\mathcal{M} \otimes \mathcal{B}(\mathcal{H})(p \otimes e_0) \cong p\mathcal{M}p \end{aligned}$$

This suggests a question that turns out to be significant: What happens if one forms the crossed product by a trace-scaling automorphism?

9.2.4 Theorem

i. [Con73] *Any factor of type III_λ $(0 < \lambda < 1)$ is isomorphic to a crossed product $\mathcal{R} \rtimes_\alpha \mathbf{Z}$, where $\mathcal{R}$ is a II_∞ factor and α is an automorphism of $\mathcal{R}$ that scales the trace by λ.*
ii. *Any factor of type III_0 is isomorphic to a crossed product $\mathcal{R} \rtimes \mathbf{Z}$, where $\mathcal{R}$ is a von Neumann algebra (not necessarily a factor) of type II_∞.*
iii. [Tak73] *Any von Neumann algebra of type III (and in particular any factor of type III_1) is isomorphic to a crossed product $\mathcal{R} \rtimes_\alpha \mathbf{R}$, where $\mathcal{R}$ is an algebra of type II_∞.*

The proof proceeds by showing that if $\mathcal{M}$ is any von Neumann algebra with modular automorphism group $\sigma : \mathbf{R} \to \mathrm{Aut}\,\mathcal{M}$, then $\mathcal{M} \rtimes_\sigma \mathbf{R}$ is semifinite. By the Takesaki duality theorem (8.7.2), we have

$$(\mathcal{M} \rtimes_\sigma \mathbf{R}) \rtimes_{\hat{\sigma}} \mathbf{R} \cong \mathcal{M} \otimes \mathcal{B}(\mathcal{H})$$

But $\mathcal{M} \otimes \mathcal{B}(\mathcal{H}) \cong \mathcal{M}$ if $\mathcal{M}$ is type III, and so $\mathcal{M}$ appears as a crossed product.

9.2.5

The existence result, Theorem 9.2.4, is accompanied by appropriate uniqueness results. For example, two crossed products $\mathcal{R}_i \rtimes_{\alpha_i} \mathbf{Z}$ as in part i are isomorphic if and only if there is an isomorphism $\mathcal{R}_1 \to \mathcal{R}_2$ such that the induced isomorphism $\mathrm{Aut}\,\mathcal{R}_1 \to \mathrm{Aut}\,\mathcal{R}_2$ carries α_1 to α_2. In particular, the Powers factors appear as crossed products with the hyperfinite II_∞ factor.

9.3 AUTOMORPHISMS OF FINITE FACTORS

In view of the results in 9.2, it is important to understand type II algebras and their automorphism groups. In this section we look at some of the ideas and results in this large and technical subject [Ped79].

9.3.1

Let $\mathcal{R}$ be a von Neumann algebra, and let $\mathcal{U}(\mathcal{R})$ be its unitary group. The map $\mathcal{U}(\mathcal{R}) \to \mathrm{Aut}\,\mathcal{R}$ taking u to $\mathrm{Ad}\,u$ is a homomorphism with kernel the center of $\mathcal{U}(\mathcal{R})$ and image the subgroup $\mathrm{Inn}\,\mathcal{R}$ of inner automorphisms. When $\mathcal{R}$ is a factor, the kernel consists of all λI with $\lambda \in \mathbf{C}$, $|\lambda| = 1$. The subgroup $\mathrm{Inn}\,\mathcal{R}$ is normal because

$$\alpha \circ \mathrm{Ad}\,u \circ \alpha^{-1} = \mathrm{Ad}\,\alpha(u) \qquad \text{for all } \alpha \in \mathrm{Aut}\,\mathcal{R}$$

and the quotient group is denoted $\mathrm{Out}\,\mathcal{R}$. An automorphism is **outer** if it is not inner. Two automorphisms of $\mathcal{R}$ are **conjugate** if they are conjugate elements of the group $\mathrm{Aut}\,\mathcal{R}$; they are **outer conjugate** if their images in $\mathrm{Out}\,\mathcal{R}$ are conjugate there (equivalently, α and β are outer conjugate if α and $\mathrm{Ad}\,u \circ \beta$ are conjugate for some u).

9.3.2

Murray and von Neumann [MvN43] introduced and studied the **fundamental group** $\mathcal{F}(\mathcal{M})$ of a type II_1 factor $\mathcal{M}$. It is the set of $\lambda \in \mathbf{R}_+$ for which there is an automorphism of $\mathcal{M} \otimes \mathcal{B}(\mathcal{H})$ that scales the trace by λ. It is a multiplicative subgroup of $\mathbf{R}_+$, and they showed that for the hyperfinite factor it is all of $\mathbf{R}_+$. Connes pursued this study and showed [Con75] that the scale factor λ is (except in the case $\lambda = 1$) a complete outer conjugacy invariant for automorphisms of the hyperfinite II_∞ factor. He also showed that automorphisms α of the hyperfinite II_1 factor $\mathcal{R}$ are classified up to outer conjugacy by the pair (p, γ), where p is the order of α in $\operatorname{Out} \mathcal{R}$ ($p = 0$ if the order is infinite), so that $\alpha^p = \operatorname{Ad} u$ for some unitary $u \in \mathcal{R}$, and $\gamma \in \mathbf{C}$ satisfies $\alpha(u) = \gamma u$ (take $\gamma = 1$ if $p = 0$). Note that such a γ is necessarily a pth root of unity; to see that γ exists, we compute

$$\alpha^{p+1}(a) = \alpha^p(\alpha(a)) = u\alpha(a)u^*$$

$$\alpha^{p+1}(a) = \alpha(\alpha^p(a)) = \alpha(uau^*) = \alpha(u)\alpha(a)\alpha(u)^*$$

for all $a \in \mathcal{R}$, from which it follows that $u^*\alpha(u)$ is in the center of $\mathcal{R}$. Connes also obtained conjugacy invariants but this problem is more difficult and is not completely solved.

9.3.3

The standard form of a II_1 factor $\mathcal{M}$ (9.1.1) may be used to study its automorphisms. Any automorphism $\alpha : \mathcal{M} \to \mathcal{M}$ has associated with it a "canonical" unitary operator u_α on $L^2(\mathcal{M})$ such that $\operatorname{Ad} u_\alpha$ leaves $\mathcal{M}$ invariant and restricts to α. In fact, because of the uniqueness of the trace, α must leave the trace invariant, and so

$$\|\alpha(a)\xi_0\|^2 = \operatorname{tr}(\alpha(a^*a)) = \operatorname{tr}(a^*a) = \|a\xi_0\|^2$$

for all $a \in \mathcal{M}$. Thus $u_\alpha(a\xi_0) = \alpha(a)\xi_0$ defines a unitary operator, and we have

$$u_\alpha a u_\alpha^* b\xi_0 = u_\alpha a\alpha^{-1}(b)\xi_0 = \alpha(a(\alpha^{-1}(b)))\xi_0 = \alpha(a)b\xi_0$$

for all $a, b \in \mathcal{M}$ so that $\alpha = \operatorname{Ad} u_\alpha$ as asserted.

We note the following additional properties of u_α:

i. If α is inner, say $\alpha = \text{Ad}\, v$ for a unitary $v \in \mathcal{M}$, then

$$u_\alpha a\xi_0 = \alpha(a)\xi_0 = vav^*\xi_0 = vJva^*\xi_0 = vJvJa\xi_0$$

and so $u_\alpha = vJvJ$. Note here that $w = JvJ \in \mathcal{M}'$; because $\mathcal{M}$ is a factor the decomposition $u_\alpha = vw$ into unitaries from $\mathcal{M}$ and $\mathcal{M}'$ is unique up to scalar multiples.

ii. u_α commutes with J:

$$u_\alpha J(a\xi_0) = u_\alpha a^*\xi_0 = \alpha(a^*)\xi_0 = \alpha(a)^*\xi_0 = J\alpha(a)\xi_0 = Ju_\alpha(a\xi_0)$$

iii. Hence $\text{Ad}\, u_\alpha$ is an automorphism of $\mathcal{M}'$:

$$u_\alpha \mathcal{M}' u_\alpha^* = u_\alpha J\mathcal{M}Ju_\alpha^* = Ju_\alpha \mathcal{M} u_\alpha^* J = J\mathcal{M}J = \mathcal{M}'$$

These facts can be used to establish a connection between certain outer automorphisms of a finite factor and certain **subfactors** (i.e., unital subalgebras that are factors):

9.3.4 Theorem

If G is a finite group of outer automorphisms of a II_1 factor $\mathcal{M}$, then

$$H \to \mathcal{M}^H = \{x \in \mathcal{M} : \alpha(x) = x \text{ for all } \alpha \in H\}$$

is a Galois correspondence between subgroups of G and subfactors of $\mathcal{M}$ that contain the fixed point subalgebra $\mathcal{M}^G$.

Sketch of Proof (see [Jon91])

If $\alpha \in G$, then the action of u_α on $\mathcal{M}'$ as above must be outer too. The canonical unitary associated with this action is u_α, again since $u_\alpha a' u_\alpha^* \xi_0 = u_\alpha a' \xi_0$ for all $a' \in \mathcal{M}'$. If this were inner, we would have $u_\alpha = vJvJ$ for some $v \in \mathcal{M}'$, and hence $\alpha = \text{Ad}\, u_\alpha = \text{Ad}(JvJ)$ would be inner on $\mathcal{M}$. Denote by $\mathcal{M}^G$ the fixed point subalgebra of the action of G on $\mathcal{M}$. Then

$$\mathcal{M}^G = \mathcal{M} \cap \{u_\alpha : \alpha \in G\}' = (\mathcal{M}' \cup \{u_\alpha : \alpha \in G\})'$$

and so

$$(\mathcal{M}^G)' = (\mathcal{M}' \cup \{u_\alpha : \alpha \in G\})'' = \mathcal{M}' \rtimes G$$

The last equality holds because the action of G on $\mathcal{M}'$ is outer. If now $\mathcal{N} \subset \mathcal{M}$ is a subfactor containing $\mathcal{M}^G$, then taking commutants gives

$$\mathcal{M}' \subset \mathcal{N}' \subset \mathcal{M}' \rtimes G$$

The proof is completed by showing that such factors $\mathcal{N}'$ are of the form $\mathcal{M}' \rtimes H$ for H a subgroup of G.

9.4 INDEX OF SUBFACTORS

Theorem 9.3.4 suggests the question: Is it possible to define an intrinsic index for subfactors that, in the situation of the theorem, corresponds via the Galois correspondence to the index of subgroups? If $[\mathcal{M} : \mathcal{N}]$ denotes the proposed index of $\mathcal{N}$ in $\mathcal{M}$, then the precise statement would be that

$$[K : H] = [\mathcal{M}^K : \mathcal{M}^H]$$

for $G \supset H \supset K$ (in particular $[\mathcal{M} : \mathcal{M}^H] = |H|$).

The answer is yes, the idea being—noting that $L^2(\mathcal{M})$ is a faithful $\mathcal{N}$-module—to construct a dimension

$$\dim_{\mathcal{N}} \mathcal{H} \in [0, +\infty]$$

for faithful $\mathcal{N}$-modules $\mathcal{H}$ and then to take

$$[\mathcal{M} : \mathcal{N}] = \dim_{\mathcal{N}} L^2(\mathcal{M})$$

9.4.1

For a finite factor $\mathcal{M}$ on $\mathcal{H}$, the quantity $\dim_{\mathcal{M}} \mathcal{H}$ is defined by taking it to be 1 when $\mathcal{H}$ is the standard representation space $L^2(\mathcal{M})$, and in general by comparing (in the sense of Murray and von Neumann) $\mathcal{H}$ with $L^2(\mathcal{M})$. To do this, choose $n \in \mathbf{N}$ so that there is an isometry

$$U : \mathcal{H} \to L^2(\mathcal{M}) \otimes \mathbf{C}^n$$

of $\mathcal{M}$-modules (if no such n exists, set $\dim_{\mathcal{M}} \mathcal{H} = +\infty$). Then UU^* is an element of $\mathcal{B}(L^2(\mathcal{M})) \otimes M_n(\mathbf{C})$; in fact, since it is a module map, it is in

$$(\mathcal{M} \otimes \mathbf{C}I_n)' = \mathcal{M}' \otimes M_n(\mathbf{C}) \cong \mathcal{M} \otimes M_n(\mathbf{C}) = M_n(\mathcal{M})$$

where the isomorphism is Ad $J \otimes \mathrm{id}$. If UU^* corresponds to $(u_{ij}) \in M_n(\mathcal{M})$, then one defines

$$\dim_{\mathcal{M}} \mathcal{H} = \sum_{i=1}^{n} \mathrm{tr}\, u_{ii}$$

As an example, take $\mathcal{M} = M_n(\mathcal{R})$ and $\mathcal{N} = \mathcal{R}^{(n)}$, where $\mathcal{R}$ is a finite factor. Then $[\mathcal{M}, \mathcal{N}] = n^2$, essentially because as vector spaces $M_n(\mathcal{R})$ is isomorphic to n^2 copies of $\mathcal{R}$.

The number $\dim_{\mathcal{M}} \mathcal{H}$ classifies the faithful representations of $\mathcal{M}$ up to unitary equivalence. It was first considered by Murray and von Neumann [MvN36], in the following form:

9.4.2 Theorem

Let $\mathcal{M}$ be a finite factor on $\mathcal{H}$ such that $\mathcal{M}'$ is finite. Fix a nonzero vector $\xi \in \mathcal{H}$, let p be the projection on the subspace $(\mathcal{M}'\xi)^-$ of $\mathcal{H}$, and let p' be the projection on $(\mathcal{M}\xi)^-$. Then $p \in \mathcal{M}$, $p' \in \mathcal{M}'$, and the ratio $c(\mathcal{M}) = \mathrm{tr}(p)/\,\mathrm{tr}(p')$ is independent of ξ.

This quantity, called the **coupling constant** of $\mathcal{M}$ on $\mathcal{H}$, measures the relative sizes of $\mathcal{M}$ and $\mathcal{M}'$. It is not difficult to see that $c_{\mathcal{M}} = \dim_{\mathcal{M}} \mathcal{H}$. For example, if $\mathcal{H} = \sum_{i=1}^{n} \oplus L^2(\mathcal{M})$ with $\mathcal{M}$ represented as the inflation $\mathcal{M}^{(n)}$, then taking $\xi = (\xi_0, 0, \ldots, 0)$, we have

$$(\mathcal{M}^{(n)}\xi)^- = L^2(\mathcal{M}) \oplus \{0\} \oplus \cdots \oplus \{0\}$$

$$((\mathcal{M}^{(n)})'\xi)^- = (M_n(\mathcal{M}')\xi)^- = \mathcal{H}$$

and so $\dim_{\mathcal{M}} \mathcal{H}$ and the coupling constant of $\mathcal{M}$ on $\mathcal{H}$ are both equal to n.

Remarks

Let $\mathcal{N}$ be a subfactor of a finite factor $\mathcal{M}$. Then

i. $[\mathcal{M} : \mathcal{N}] < \infty$ if and only if the commutant of $\mathcal{N}$, computed on $L^2(\mathcal{M})$, is finite.

ii. $[\mathcal{M} : \mathcal{N}] = [\mathcal{N}' : \mathcal{M}']$.

9.5 INDEX VALUES

For a fixed finite type II factor $\mathcal{M}$, what can be said about the possible values of $[\mathcal{M} : \mathcal{N}]$? One always has the values $\{1, 4, 9, \ldots\} \cup \{+\infty\}$, and it is conceivable that there could be a factor with only these values. At the other extreme, could one get all of $[1, +\infty]$ for some $\mathcal{M}$?

9.5.1 *Theorem* (Jones)

Let $\mathcal{M}$ be a finite type II factor.

i. *If $\mathcal{N}$ is a subfactor* (*with the same identity*) *and $[\mathcal{M} : \mathcal{N}] < 4$, then*

$$[\mathcal{M} : \mathcal{N}] = 4\cos^2 \frac{\pi}{n} \qquad \textit{for some } n = 3, 4, \ldots$$

ii. *If $\mathcal{M}$ is the hyperfinite II_1 factor then*

$$\{[\mathcal{M} : \mathcal{N}] \colon \mathcal{N} \textit{ subfactor}\} = \left\{4\cos^2\left(\frac{\pi}{n}\right) : n \geq 3\right\} \cup [4, +\infty]$$

Thus the first few allowed values of the index are $1, 2, (3+\sqrt{5})/2, 3, \ldots,$ and this sequence increases to 4.

9.5.2

The main ingredient of the proof is what Jones called the **basic construction**. Take $\mathcal{M}$ in its standard representation on $L^2(\mathcal{M})$ (9.1.1), and let $\mathcal{N}$ be a subfactor. From uniqueness of the trace, it follows that the trace on $\mathcal{N}$ is the restriction of the trace on $\mathcal{M}$ and hence that $L^2(\mathcal{N})$ can be identified as a subspace of $L^2(\mathcal{M})$. Let e be the projection of $L^2(\mathcal{M})$ onto $L^2(\mathcal{N})$. It has the following properties:

1. $\mathcal{M} \cap \{e\}' = \mathcal{N}$.
2. The von Neumann algebra $\langle \mathcal{M}, e\rangle = (\mathcal{M} \cup \{e\})''$ is a factor, finite if $\mathcal{N}'$ is (equivalently, by remark i of 9.4.2, if $[\mathcal{M} : \mathcal{N}] < \infty$).
3. $[\langle \mathcal{M}, e\rangle : \mathcal{M}] = [\mathcal{M} : \mathcal{N}]$.
4. $\mathrm{tr}(ae) = \mathrm{tr}(a)/[\mathcal{M} : \mathcal{N}]$ for all $a \in \mathcal{M}$.

In property 4, tr denotes the trace on $\langle \mathcal{M}, e\rangle$ (which exists by property 2); by uniqueness it restricts to the trace on $\mathcal{M}$. To prove these assertions, we need to know that e is the continuous extension of a conditional expectation $\epsilon : \mathcal{M} \to \mathcal{N}$.

9.5.3 *Definition*

A **conditional expectation** of a von Neumann algebra $\mathcal{M}$ on a subalgebra $\mathcal{N}$ is a unital positive linear map $\epsilon : \mathcal{M} \to \mathcal{N}$ such that

$$\epsilon(ab) = \epsilon(a)b \qquad \text{for all } a \in \mathcal{M},\ b \in \mathcal{N}$$

It follows easily that ϵ has these properties:

i. $\epsilon(a)^* = \epsilon(a^*)$ for all $a \in \mathcal{M}$.
ii. $\epsilon(ba) = b\epsilon(a)$ for all $a \in \mathcal{M}, b \in \mathcal{N}$.
iii. $\epsilon \circ \epsilon = \epsilon$.
iv. $\epsilon(a)^*\epsilon(a) \leq \epsilon(a^*a)$ for all $a \in \mathcal{M}$.
v. $\|\epsilon\| = 1$.

For property i use the positivity of ϵ, for property ii take adjoints in the defining equation and use property i, for property iii replace a by I and b by $\epsilon(a)$ in property ii, for property iv expand $\epsilon((a - \epsilon(a))^*(a - \epsilon(a)) \geq 0$, and for property v use property iv and $a^*a \leq \|a\|^2 I$.

Examples

As examples of conditional expectations, we mention the center-valued trace $\mathcal{M} \to \mathcal{Z}(\mathcal{M})$ of 4.8.3 and any state $\mathcal{M} \to \mathbf{C}$. If $\{p_i\}$ is an orthogonal family of projections in $\mathcal{M}$ with $\sum p_i = I$, then $\epsilon(A) = \sum p_i a p_i$ defines a conditional expectation $\epsilon : \mathcal{M} \to \mathcal{M} \cap \{p_i\}'$.

9.5.4

This terminology is suggested by the picture of a von Neumann algebra as a noncommutative measure space (or rather an L^∞ space). In our situation—a subfactor $\mathcal{N}$ of a finite type II factor $\mathcal{M}$—the existence of the conditional expectation $\epsilon : \mathcal{M} \to \mathcal{N}$ comes from the same picture, via the following noncommutative **Radon-Nikodym property** [Sak71, 1.24.3]. The normal positive linear functionals φ on $\mathcal{N}$ correspond, via the trace, to the positive elements h of $\mathcal{N}$:

$$\varphi(x) = \operatorname{tr}(hx) \qquad \text{for all } x \in \mathcal{N}$$

It follows that for any $a \in \mathcal{M}$ there exists $\epsilon(a) \in \mathcal{N}$ such that

$$\operatorname{tr}(ax) = \operatorname{tr}(\epsilon(a)x) \qquad \text{for all } x \in \mathcal{N} \tag{$*$}$$

Moreover $\epsilon(a)$ is unique (because the trace is faithful), and it is easy to see that $\epsilon : \mathcal{M} \to \mathcal{N}$ is a conditional expectation.

9.5.5

The connection between ϵ and e (the projection of $L^2(\mathcal{M})$ on $L^2(\mathcal{N})$) is now immediate. Recalling that $L^2(\mathcal{M})$ is the GNS representation space for the trace on $\mathcal{M}$, we see that equation $(*)$ is just the assertion that

$$a - \epsilon(a) \perp L^2(\mathcal{N}) \qquad \text{for all } a \in \mathcal{M}$$

and hence $e(a\xi_0) = \epsilon(a)\xi_0$ for all $a \in \mathcal{M}$.

Proof of 9.5.2, Properties 1–4

(1) If $b \in \mathcal{N}$, then for all $a \in \mathcal{M}$ we have

$$eb(a\xi_0) = \epsilon(ba)\xi_0 = b\epsilon(a)\xi_0 = be(a\xi_0)$$

and so $b \in \{e\}'$. If $a \in \mathcal{M} \cap \{e\}'$, then

$$\epsilon(a)\xi_0 = ea\xi_0 = ae\xi_0 = a\xi_0$$

so that $a = \epsilon(a) \in \mathcal{N}$.

(2) Note first that $J \in \{e\}'$, since for all $a \in \mathcal{M}$

$$eJ(a\xi_0) = ea^*\xi_0 = \epsilon(a^*)\xi_0 = \epsilon(a)^*\xi_0 = J\epsilon(a)\xi_0 = Je(a\xi_0)$$

We then have successively

$$J\langle\mathcal{M}, e\rangle J = \langle J\mathcal{M}J, JeJ\rangle = \langle\mathcal{M}', e\rangle$$

$$(J\langle\mathcal{M}, e\rangle J)' = \mathcal{M} \cap \{e\}' = \mathcal{N}$$

$$\langle\mathcal{M}, e\rangle = J\mathcal{N}'J$$

(3) By remark ii of 9.4.2 and the proof just given

$$[\langle\mathcal{M}, e\rangle : \mathcal{M}] = [\mathcal{M}' : \langle\mathcal{M}, e\rangle'] = [\mathcal{M}' : J\mathcal{N}J]$$
$$= [J\mathcal{M}'J : \mathcal{N}] = [\mathcal{M} : \mathcal{N}]$$

(4) We are assuming that $[\mathcal{M} : \mathcal{N}] < \infty$, so that by property 2, $\langle\mathcal{M}, e\rangle$ is a finite factor; by uniqueness its trace extends the trace on $\mathcal{M}$. By property 1, $b \to \mathrm{tr}(eb)$ is a trace on $\mathcal{N}$ (not necessarily normalized), so there exists $\tau > 0$ (namely $\tau = \mathrm{tr}(e)$) such that

$$\mathrm{tr}(eb) = \tau(\mathrm{tr}(b)) \qquad \text{for all } b \in \mathcal{N}$$

Now for any $a, a_1 \in \mathcal{M}$, we have

$$eae(a_1\xi_0) = \epsilon(a\epsilon(a_1))\xi_0 = \epsilon(a)\epsilon(a_1)\xi_0 = \epsilon(a)e(a_1\xi_0)$$

so that $eae = \epsilon(a)e$. Hence

$$\mathrm{tr}(ae) = \mathrm{tr}(eae) = \mathrm{tr}(\epsilon(a)e) = \tau(\mathrm{tr}(\epsilon(a))) = \tau(\mathrm{tr}(a))$$

where we have used $(*)$ with $x = I$. Finally, applying 9.4.2 to $\mathcal{N}$ and ξ_0 and noting that $(\mathcal{N}'\xi_0)^- = L^2(\mathcal{M})$ and $(\mathcal{N}\xi_0)^- = L^2(\mathcal{N})$, we have

$$[\mathcal{M} : \mathcal{N}] = \frac{\mathrm{tr}(I)}{\mathrm{tr}(e)} = \tau^{-1}$$

This completes the basic construction 9.5.2.

9.5.6

Let $\mathcal{M}$ be a finite type II factor and $\mathcal{N} \subset \mathcal{M}$ a subfactor of finite index. By applying the basic construction, we obtain a new factor-subfactor pair $\mathcal{M} \subset \langle\mathcal{M}, e\rangle$ of the same index, by properties 2 and 3 of 9.5.2. Hence the basic construction can be repeated. Iteration produces a tower

$$\mathcal{N} \subset \mathcal{M} \subset \langle\mathcal{M}, e_1\rangle \subset \langle\langle\mathcal{M}, e_1\rangle, e_2\rangle \subset \cdots$$

of finite type II factors. The sequence $e_1, e_2, \ldots$ of **Jones projections** satisfies the following relations:

1. $e_i e_{i\pm1} e_i = \tau e_i$.

2. $e_i e_j = e_j e_i$ if $|i - j| \geq 2$.
3. $\mathrm{tr}(e_n x) = \tau\, \mathrm{tr}(x)$ for all $x \in \langle \mathcal{M}, e_1, \ldots, e_{n-1} \rangle$.

The second of these follows from property 1 of 9.5.2. Notice that because of property 3 (9.5.2), the iteration produces the same value of τ at each stage; hence equation 3 above results from property 4 (9.5.2). From the proof of property 4 (9.5.2) we have $e_i e_{i-1} e_i = \epsilon_i(e_{i-1}) e_i$. Also

$$(\epsilon_i(e_{i-1})e_i - \tau e_i)^2 = e_i \epsilon_i(e_{i-1})^2 e_i - 2\tau e_i \epsilon_i(e_{i-1})e_i + \tau^2 e_i$$
$$= \epsilon_i(\epsilon_i(e_{i-1})e_{i-1})e_i - 2\tau \epsilon_i(e_{i-1})e_i + \tau^2 e_i$$

since e_i and $\epsilon_i(e_{i-1})$ commute by property 1 of 9.5.2, and this has trace 0 by property 4 and the fact that ϵ_i leaves the trace invariant. Hence

$$\epsilon_i(e_{i-1})e_i - \tau e_i = 0 \quad \text{and} \quad e_i e_{i-1} e_i = \tau e_i$$

Finally, straightforward computation gives

$$\mathrm{tr}(e_{i-1} e_i e_{i-1} - \tau e_{i-1})^2 = \mathrm{tr}(e_i e_{i-1} e_i - \tau e_i)^2 = 0$$

and therefore $e_{i-1} e_i e_{i-1} = \tau e_{i-1}$.

9.5.7

The proof of Jones's Theorem 9.5.1(i) is now completed by showing that equations 1–3 (9.5.6) force $[\mathcal{M} : \mathcal{N}] = \tau^{-1}$ to satisfy the stated restriction. Let $c_n = 4\cos^2(\pi/n)$ for $n \geq 3$. Define a sequence of polynomials by

$$p_0 = p_1 = 1 \quad \text{and} \quad p_{n+2}(x) = p_{n+1}(x) - x p_n(x), \qquad n \geq 0$$

Then for $n \geq 2$ (and $1/c_2 = +\infty$)

$$p_n > 0 \qquad \text{on } (0, 1/c_{n+1})$$
$$p_n < 0 \qquad \text{on } (1/c_{n+1}, 1/c_n)$$

Let us suppose, if possible, that τ is in this last interval for some n. Define operators $f_0, f_1, \ldots, f_{n-1}$ on $L^2(\mathcal{M})$ recursively by $f_0 = I$ and

$$f_k = f_{k-1} - \left(\frac{p_{k-1}(\tau)}{p_k(\tau)} \right) f_{k-1} e_k f_{k-1}, \qquad 1 \leq k \leq n-1$$

Evidently f_k is in the subalgebra generated by $e_1, \ldots, e_k$, and so f_k commutes with e_{k+2} by equation 2 of 9.5.6.

Lemma

The f_k are projections and for all $k \geq 0$ we have

$$(e_{k+1} f_k)^2 = \left(\frac{p_{k+1}(\tau)}{p_k(\tau)} \right) e_{k+1} f_k$$

The proof of the lemma is by induction. The assertions are trivially true for $k = 0$. Assume them true for a given value of k. Then straightforward computation using only the hypotheses shows that f_{k+1} is a projection. From the definition of f_{k+1}, we have

$$e_{k+2} f_{k+1} = e_{k+2} f_k - \left(\frac{p_k(\tau)}{p_{k+1}(\tau)} \right) e_{k+2} f_k e_{k+1} f_k$$

Squaring this expression and using that e_{k+2} and f_k are commuting projections, that $e_{k+2}e_{k+1}e_{k+2} = \tau e_{k+2}$ (from equation 1, 9.5.6) and that

$$1 - \frac{\tau p_k(\tau)}{p_{k+1}(\tau)} = \frac{p_{k+2}(\tau)}{p_{k+1}(\tau)}$$

gives the second statement.

The final step in the proof of 9.5.1(i) is to use equation 3, 9.5.6, to compute the trace of the f_k. We have

$$\begin{aligned} \operatorname{tr}(f_k) &= \operatorname{tr}(f_{k-1}) - \left(\frac{p_{k-1}(\tau)}{p_k(\tau)} \right) \tau(\operatorname{tr}(f_{k-1})) \\ &= \left(\frac{p_{k+1}(\tau)}{p_k(\tau)} \right) \operatorname{tr}(f_{k-1}) \end{aligned}$$

and $\operatorname{tr}(f_0) = 1$, so

$$\operatorname{tr}(f_{n-1}) = p_n(\tau) < 0$$

by our assumption about τ. But f_{n-1} is a projection, and we have arrived at a contradiction.

This proof is a somewhat rearranged and fleshed-out version of the argument in [Jon91]. There it is pointed out that the projection f_k is just the complement of the supremum of $\{e_1, e_2, \ldots, e_k\}$. If $\tau = 1/c_{n+1}$ so that $p_n(\tau) = 0$, then $f_{n-1} = 0$ and $\{e_1, e_2, \cdots, e_{n-1}\}$ has supremum 1.

9.5.8 *Examples: Discrete Index Values*

Given a finite type II factor $\mathcal{M}$ and a subfactor $\mathcal{N}$ with $[\mathcal{M} : \mathcal{N}] < \infty$, let

$$\mathcal{N} \subset \mathcal{M} \subset \mathcal{M}_1 \subset \mathcal{M}_2 \subset \cdots$$

be the tower obtained by iterating the basic construction. The union has a unique trace and the von Neumann algebra $\mathcal{M}_\infty$ it generates (on $L^2(\mathcal{M})$) is a finite type II factor. Moreover the subalgebras

$$\mathcal{S} = \{e_1, e_2, \ldots\}'' \quad \text{and} \quad \mathcal{T} = \{e_2, e_3, \ldots\}''$$

are subfactors with $[\mathcal{S} : \mathcal{T}] = \tau^{-1}$. In fact this will be true for any sequence $\{e_i\}$ of projections in a finite type II factor that satisfy the equations 1–3 of 9.5.6 (delete $\mathcal{M}$ in equation 3). Such sequences can be constructed by starting with a suitable inclusion $B \subset A$ of finite-dimensional algebras and iterating the basic construction. The main difficulty is in choosing the weightings of the traces on the simple summands in order to obtain an appropriate trace on A; this can be done (uniquely) provided that the Bratteli diagram of the inclusion is connected [Jon83]. In particular, subfactors with the discrete series of index values can be constructed in this way.

The Bratteli diagrams of towers like these can easily be written down. From the proof of 9.5.2, property 2, we have that the centers of B and $\langle A, e\rangle$ have the same dimension, and in fact it can be shown that the diagram of the inclusion $A \subset \langle A, e\rangle$ is just the mirror image of the starting inclusion $B \subset A$.

9.5.9 *Examples: Index Values* ≥ 4

Let $\mathcal{R}$ be the hyperfinite II_1 factor, and recall (9.3.2) that its fundamental group is $\mathbf{R}_+$. This means (9.2.3) that for any projection $p \in \mathcal{R}$, $p \neq 0$, we have $p\mathcal{R}p \cong \mathcal{R}$. Hence for any projection $p \in \mathcal{R}$ with $p \neq 0, 1$, there is an isomorphism $\alpha : p\mathcal{R}p \to p^\perp \mathcal{R} p^\perp$. Then

$$\mathcal{N} = \{x + \alpha(x) : x \in p\mathcal{R}p\}$$

is a subfactor and

$$[\mathcal{R} : \mathcal{N}] = \frac{1}{t} + \frac{1}{1-t}$$

where $t = \operatorname{tr}(p)$. Before indicating how this can be proved, we note that because p is arbitrary t runs over all values in (0, 1), and hence that $[\mathcal{R} : \mathcal{N}]$ will run over all of $[4, +\infty)$.

Let $\mathcal{M}$ be any von Neumann algebra on a Hilbert space $\mathcal{H}$, and let p be a projection in $\mathcal{M}$. We will need certain facts about the algebras $p\mathcal{M}p$ and $\mathcal{M}'p$, regarded as algebras of operators on $p\mathcal{H}$:

i. $p\mathcal{M}p$ is a von Neumann algebra with commutant $\mathcal{M}'p$.
ii. $\mathcal{Z}(p\mathcal{M}p) = \mathcal{Z}(\mathcal{M})p$, where $\mathcal{Z}$ denotes the center; in particular, if $\mathcal{M}$ is a factor, so is $p\mathcal{M}p$ [Dix81, 1.2.1].

From these it follows that $\mathcal{N}$ (defined above) is a factor. Note also that $\mathcal{N} \cong p\mathcal{R}p$ and $\mathcal{N}' \cong p\mathcal{R}'p$ are finite and that $p \in \mathcal{N}'$. To compute $[\mathcal{R} : \mathcal{N}]$, we need the following relation between coupling constants (9.4.2):

iii. If $\mathcal{N}$ is a finite factor with finite commutant $\mathcal{N}'$ and p is a projection in $\mathcal{N}'$, then $c(\mathcal{N}p) = \operatorname{tr}'(p)c(\mathcal{N})$, where tr' denotes the trace on $\mathcal{N}'$ [Dix81, 3.6.2].

Applying this to our situation we have

$$\dim_{\mathcal{N}p}(p\mathcal{H}) = c(\mathcal{N}p) = \operatorname{tr}'(p)c(\mathcal{N}) = \operatorname{tr}'(p)\dim_{\mathcal{N}}\mathcal{H}$$

Taking $\mathcal{H} = L^2(\mathcal{R})$ and noting that $\mathcal{N}p = p\mathcal{R}p$, this gives

$$\dim_{p\mathcal{R}p}(pL^2(\mathcal{R})) = \operatorname{tr}'(p)[\mathcal{R} : \mathcal{N}]$$

It is not difficult to see that the left-hand side of this is just $1/\operatorname{tr}(p)$, the trace being computed in $\mathcal{R}$. For example, if $\operatorname{tr}(p) = 1/n$, then $\mathcal{R} \cong M_n(p\mathcal{R}p)$ and $pL^2(\mathcal{R})$ can be identified as n copies of $L^2(p\mathcal{R}p)$. Hence

$$\operatorname{tr}(p)\operatorname{tr}'(p)[\mathcal{R} : \mathcal{N}] = 1$$

The same is true with p replaced by $p^\perp$ and so

$$\frac{1}{t} + \frac{1}{1-t} = (\operatorname{tr}'(p) + \operatorname{tr}'(p^\perp))[\mathcal{R} : \mathcal{N}] = [\mathcal{R} : \mathcal{N}]$$

Remarks

The discrete series of index values occurs elsewhere (see [GHJ89]). In the above we had $p \in \mathcal{R} \cap \mathcal{N}'$, and this raises the question of which index values are possible in the ("irreducible") case $\mathcal{M} \cap \mathcal{N}' = \mathbf{C}I$. It has been conjectured that $[\mathcal{M} : \mathcal{N}]$ must be algebraic.

The study of subfactors continues to be intensive, and a great deal more is known about them than we have been able to describe. For more information the reader should consult Popa's forthcoming CBMS lectures [Pop95].

9.6 BRAIDS AND KNOTS

In this section we will describe the connection between subfactors and knot theory. This connection has inspired a great deal of recent work by many people and has led to further connections with physics and the topology of low-dimensional manifolds, as well as earning its discoverer, Vaughan Jones, a Fields Medal in 1990. For more details (and a good sample of his inimitable style) the reader should consult Jones's CBMS monograph [Jon91].

9.6.1

The ***n*-string braid group** B_n is defined in terms of the plane $\mathbf{R}^2$ in which n points have been distinguished. Consider two horizontal copies in $\mathbf{R}^3$ of such a plane, together with n nonintersecting smooth paths ("strings") joining the distinguished points of the two planes in pairs. To avoid such complications as knots in the individual strings, we require that the tangent vector is never parallel to the planes. The elements of B_n ("n-string braids") are then equivalence classes of such objects, two of them being regarded as equivalent if each can be deformed into the other without intersections or horizontal tangents on the way. The group operation comes from concatenation (deleting the middle plane), so that the identity is the braid with vertical strings and the inverse of a braid is its mirror image:

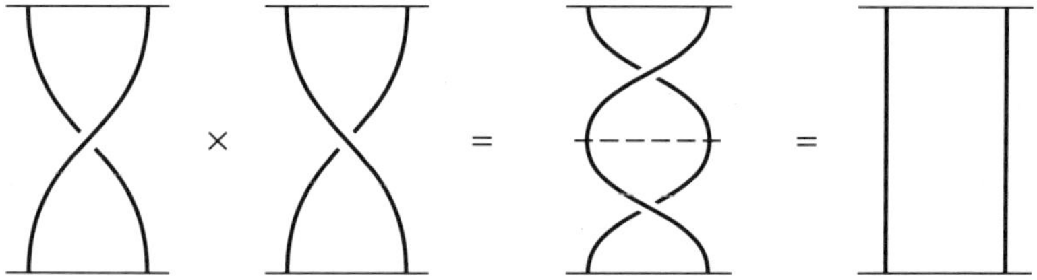

9.6.2

Emil Artin gave the following presentation of the braid group B_n [Art47]:

$$\text{Generators:} \quad \sigma_1, \sigma_2, \ldots, \sigma_{n-1}$$

$$\text{Relations:} \quad \begin{cases} \sigma_i\sigma_j = \sigma_j\sigma_i & \text{if } |i - j| \geq 2 \\ \sigma_i\sigma_{i+1}\sigma_i = \sigma_{i+1}\sigma_i\sigma_{i+1} \end{cases}$$

Here σ_i is the braid pictured above on the left, considered as connecting points i and $i + 1$ to each other and all other points vertically.

9.6.3

We have seen relations like these before, namely in 9.5.6, and this suggests looking for representations of the braid group in the von Neumann algebra $\langle 1, e_1, e_2, \cdots \rangle$ generated by the projections e_i in the tower of subfactors obtained by iterating the basic construction. If e is a projection and $a, b \in \mathbf{R} \setminus \{0\}$, the operator $ae + be^\perp$ is invertible and hence is a candidate for the image of a generator. Since the relations are homogeneous, we take $b = -1$ and try

$$\sigma_i \to te_i - e_i^\perp = (t + 1)e_i - 1$$

with t real and nonzero. The first relation is automatically satisfied because of equation 9.5.6(2); using equation 9.5.6(1) the second gives

$$((t + 1)^2\tau - t)e_i = ((t + 1)^2\tau - t)e_{i+1}$$

or $\tau^{-1} = 2 + t + t^{-1}$. This determines a braid group representation π.

9.6.4

What is the significance of equation 9.5.6(3) in the present context? This seemingly naive question has a remarkably satisfying and significant answer.

A **knot** is a smoothly embedded circle in $\mathbf{R}^3$, and a **link** is a finite disjoint union of knots. There is a close connection between braids and links: From any braid $\beta \in B_n$ one obtains a link $\hat{\beta}$ by connecting each end point at the top of β with the end point directly below it, and moreover every link arises in this

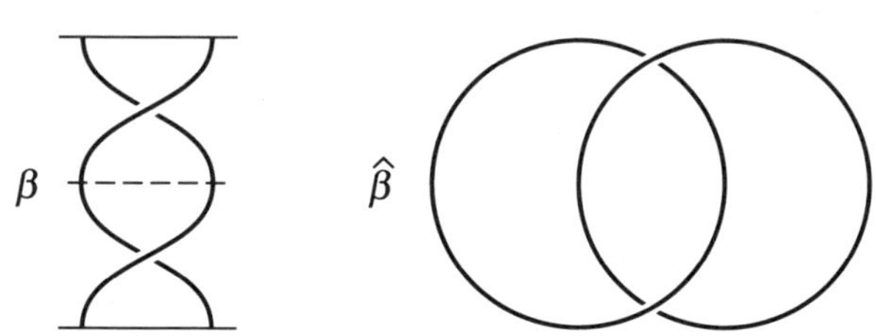

way. Is a link determined by a unique braid? It is not difficult to see, for any $\alpha, \beta \in B_n$, that β and $\alpha\beta\alpha^{-1}$ give the same link (or, equivalently, that $\alpha\beta$ and

$\beta\alpha$ give the same link). This is interesting, since any "trace" would make the same identification. But there is more: for any $\beta \in B_n$ the braids β and $\beta'\sigma_n^{\pm 1}$ give the same link, where $\beta' \in B_{n+1}$ is obtained from β by adding a vertical string.

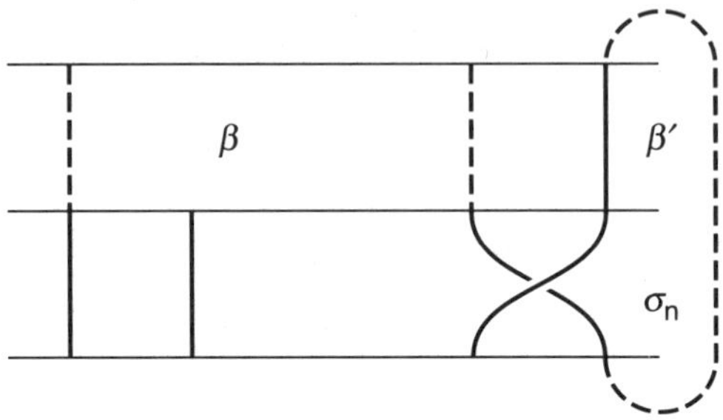

Markov's Theorem [Mar35]

Any two braids $\alpha \in B_m$ and $\beta \in B_n$ that give the same link are connected by a sequence of transformations of the above two types.

We can now begin to see the significance of equation 9.5.6(3). For all n and all $\beta \in B_n$, we compute (with π as in 9.6.3)

$$\begin{aligned}\operatorname{tr}(\pi(\beta'\sigma_n)) &= \operatorname{tr}(\pi(\beta)((t+1)e_n - 1))\\ &= ((t+1)\tau - 1)\operatorname{tr}(\pi(\beta))\\ &= -\frac{1}{t+1}\operatorname{tr}(\pi(\beta))\end{aligned}$$

and similarly $\operatorname{tr}(\pi(\beta'\sigma_n^{-1})) = (-t/(t+1))\operatorname{tr}(\pi(\beta))$, making use of

$$((t+1)e_n - 1)^{-1} = (te_n - e_n^{\perp})^{-1} = \frac{1}{t}e_n - e_n^{\perp} = \left(\frac{t+1}{t}\right)e_n - 1$$

This suggests the possibility of "normalizing" the quantity $\operatorname{tr}(\pi(\beta))$ in such a way as to get the same value for β and $\beta\sigma_n^{\pm 1}$. For any braid $\beta \in B_n$, define

$$V_\beta(t) = \left(-\frac{t+1}{\sqrt{t}}\right)^{n-1}\left(\sqrt{t}\right)^{e(\beta)}\operatorname{tr}(\pi(\beta))$$

where $e(\beta)$ is the sum of all the exponents in any representation of β as a product of generators and their inverses. Note that $e(\beta)$ is well-defined because the defining relations for B_n are homogeneous. Now using the computations of $\mathrm{tr}(\pi(\beta'\sigma_n^{\pm 1}))$ and noting that

$$e(\beta'\sigma_n^{\pm 1}) = e(\beta) \pm 1$$

one easily sees that $V_{\beta'\sigma_n^{\pm 1}} = V_\beta$ for all $\beta \in B_n$.

Since it is clear that V is constant on conjugacy classes in each B_n, we see by Markov's theorem that for any $\beta \in \cup B_n$ the polynomial V_β only depends on the link $\hat{\beta}$. We therefore have a link invariant, known as the **Jones polynomial** of the link. It is clearly a polynomial in $\sqrt{t}$ and $1/\sqrt{t}$.

9.6.5

Here are some examples of knot polynomials.

EXAMPLE 1. We have $\sigma_1 \in B_2$ and

$$\mathrm{tr}(\pi(\sigma_1)) = \mathrm{tr}((t+1)e_1 - I) = (t+1)\tau - 1 = -\frac{1}{t+1}$$

and therefore

$$V_{\sigma_1} = \left(-\frac{t+1}{\sqrt{t}}\right)\sqrt{t}\,\mathrm{tr}(\pi(\sigma_1)) = 1$$

Here $\widehat{\sigma_1}$ is the "unknot", an unknotted circle; hence the Jones polynomial of the unknot is 1.

EXAMPLE 2. We can regard σ_1 as an element of B_3 by adding a vertical string. Then $\widehat{\sigma_1}$ is two unlinked circles, and the Jones polynomial is $-(t+1)/\sqrt{t}$.

EXAMPLE 3. For $\sigma_1^2 \in B_2$ we have that $\widehat{\sigma_1^2}$ is two linked circles and that

$$\pi(\sigma_1^2) = (\pi(\sigma_1))^2 = (te_1 - e_1^\perp)^2 = t^2 e_1 + e_1^\perp = (t^2-1)e + 1$$

$$\mathrm{tr}(\pi(\sigma_1^2)) = (t^2-1)\tau + 1 = \frac{(t^2-1)t}{(t+1)^2} + 1 = \frac{t^2+1}{t+1}$$

$$V_{\sigma^2} = \left(-\frac{t+1}{\sqrt{t}}\right)(\sqrt{t})^2\left(\frac{t^2+1}{t+1}\right) = -\sqrt{t}(t^2+1)$$

EXAMPLE 4. For $\sigma_1^3 \in B_2$ the corresponding link is a trefoil knot—a circle with a single overhand knot in it. We compute

$$\pi(\sigma_1^3) = (te_1 - e_1^\perp)^3 = t^3 e_1 - e_1^\perp = (t^3+1)e_1 - 1$$

$$\mathrm{tr}(\pi(\sigma_1^3)) = (t^3+1)\left(\frac{t}{(t^2+1)^2}\right) - 1 = \frac{t^3 - t^2 - 1}{t+1}$$

$$V_{\sigma_1^3} = \left(-\frac{t+1}{\sqrt{t}}\right)(\sqrt{t})^3\left(\frac{t^3-t^2-1}{t+1}\right) = -t^4 + t^3 + t$$

EXAMPLE 5. The trefoil knot has two possible orientations, the other coming from $\sigma_1^{-3} \in B_2$. We have

$$\pi(\sigma_1^{-3}) = \left(\frac{1}{t}e_1 - e_1^\perp\right)^3 = \left(\frac{1}{t^3} + 1\right)e_1 - 1$$

$$\mathrm{tr}(\pi(\sigma_1^{-3})) = \left(\frac{t^3+1}{t^3}\right)\left(\frac{t}{(t+1)^2}\right) - 1 = \frac{1 - t - t^3}{t^2(t+1)}$$

$$V_{\sigma_1^{-3}} = \left(-\frac{t+1}{\sqrt{t}}\right)(\sqrt{t})^{-3}\left(\frac{1-t-t^3}{t^2(t+1)}\right) = \frac{t^3+t-1}{t^4}$$

Observe that substituting $1/t$ for t in this result produces the Jones polynomials for $\sigma_1^3 \in B_2$. In fact it is true generally that

$$V_{\beta^{-1}}\left(\frac{1}{t}\right) = V_\beta(t)$$

for any braid β.

Chapter Ten

K-Theory

10.1 INTRODUCTION

The origin of K-theory is usually traced to Grothendieck's 1958 generalization of the Riemann-Roch theorem. The classical theorem gives a formula for the number of linearly independent meromorphic functions with prescribed poles and multiplicities on a compact Riemann surface. Here the basic quantity is an integer, the dimension of a certain finite-dimensional vector space. For generalizations one needs a more flexible way of measuring or counting, and Grothendieck's idea was to replace the integers by more general abelian groups. This idea was already present in Murray and von Neumann's famous 1936 paper *On Rings of Operators* [MvN36], where a real-valued dimension function was found to be appropriate. They worked with abelian semigroups, specifically the dimension semigroup $D(M) = P(M)/\sim$ of a von Neumann algebra M. Grothendieck, however, passed to the enveloping group of the abelian semigroup (isomorphism classes of coherent sheaves over an algebraic variety) that appeared in his work. In this process information is lost, but the result is a much more computable object.

Topological K-theory was developed shortly thereafter by Atiyah and Hirzebruch [AtH61], following Grothendieck's line of development but with vector bundles over a compact space in place of sheaves. In fact they showed that much more is possible in this setting. By making use of the topological notion of suspension and Bott's fundamental periodicity result [Bot59], they constructed a generalized cohomology theory of great power and beauty. This work reached a culmination with the Atiyah-Singer index theorem [AtS63; AtS68].

It was soon realized that much of the theory, suitably formulated, extends to general Banach algebras. For C^*-algebras the resulting theory, **operator *K*-theory**, is especially nice: It is characterized (more or less) by a simple and elegant set of axioms, it admits various generalizations (due to Connes and others) of the index theorem, and it sheds light on certain problems in topology itself.

The basic reference for operator algebra K-theory is [Bla86]. For a more leisurely presentation [Weg93] can be recommended (and it comes with a bonus of Danish good humour and whimsicality). Good introductions are included in [Mur90] and [Dav95]. For topological K-theory see [Ati67].

10.2 TOPOLOGICAL K-THEORY

Topological K-theory is a generalized cohomology theory on the category of compact Hausdorff spaces. This means that it satisfies the Eilenberg-Steenrod axioms [EiS52] except for the dimension axiom. It is based on a functor, K, from compact Hausdorff spaces to abelian groups which we now construct.

10.2.1

The first step is to form, for a compact Hausdorff space X, the set $V(X)$ of isomorphism classes of complex vector bundles over X. This has the structure of an abelian semigroup with identity, where the operation is defined by means of direct sum (see 3.4). Furthermore V is a contravariant functor, in the sense that a continuous map $f: X \to Y$ of compact Hausdorff spaces induces a homomorphism $f: V(Y) \to V(X)$ of semigroups in a way that respects composition of maps. This is defined by means of the **pullback** construction: a bundle $p: E \to Y$ over Y induces the bundle f^*E over X given by

$$f^*E = \{(x, v) \in X \times E : f(x) = p(v)\}$$

with the obvious projection $f^*E \to X$.

10.2.2

It should be noted that $V(X)$ need not have cancellation. That is, it is possible to have vector bundles E, F, G with $E \oplus F \cong E \oplus G$ but $F \ncong G$. This can occur even for compact manifolds, and it is possible for torsion to be present, in the sense that a finite direct sum of copies of a nontrivial bundle can be trivial. A general question, usually too difficult to answer, is to describe the structure of these semigroups [Rie87; Bla86]. Note, however, that for X a singleton we have $V(X) \cong \mathbf{N}$ via dimension.

10.2.3

To any abelian semigroup V one can associate, in a canonical way, an abelian group $K(V)$—the **Grothendieck** or **enveloping group** of V—and a homomorphism $\varphi : V \to K(V)$ with the property that any homomorphism $V \to A$, A an abelian group, factors uniquely through φ. It is clear from this property that the pair $(K(V), \varphi)$ is unique up to isomorphism. For existence, define an equivalence relation on $V \times V$ by

$$(v_1, v_1') \sim (v_2, v_2') \text{ if and only if } v_1 + v_2' + v = v_1' + v_2 + v$$

for some $v \in V$, and take $K(V)$ to consist of the equivalence classes and $\varphi(v)$ to be the equivalence class of $(v + v_0, v_0)$ for any fixed $v_0 \in V$. From the universal property it is immediate that any homomorphism $V_1 \to V_2$ of abelian semigroups fits into a commutative square

$$\begin{array}{ccc} V_1 & \longrightarrow & V_2 \\ \downarrow & & \downarrow \\ K(V_1) & \longrightarrow & K(V_2) \end{array}$$

Thus $V \to K(V)$ is a covariant functor.

10.2.4

The **topological *K*-functor** is now defined, for any compact Hausdorff space X, by $K(X) = K(V(X))$. Topological K-theory is based on this contravariant functor; it provides a conceptual framework that facilitates computations and applications. How this is done is indicated below, in the more general case of Banach algebras.

10.2.5

At this point one can make very direct contact with Hilbert space operators. Let $\mathcal{F}$ be the space of Fredholm operators on a complex separable Hilbert space (see [Dou72] for a good account of these), and let $\mathcal{F}$ have the operator norm topology. Since the product (= composition) of Fredholm operators is again Fredholm, it follows that for any topological space X the space $[X, \mathcal{F}]$ of homotopy classes of continuous maps $X \to \mathcal{F}$ is a group.

Theorem [Ati67]; [Jän65]

For compact Hausdorff spaces X there are natural isomorphisms

$$\text{Index} : [X, \mathcal{F}] \to K(X)$$

The reason for calling this map an "index" can be understood by considering what happens when X is a single point. Then $[X, \mathcal{F}]$ is identified with the path-components of $\mathcal{F}$, and the latter are just the sets where the index map $\mathcal{F} \to \mathbf{Z}$ is constant. On the other hand, it was noted in 10.2.2 that $V(\text{point}) \cong \mathbf{N}$ and so $K(\text{point}) \cong K(\mathbf{N}) = \mathbf{Z}$.

If $F : X \to \mathcal{F}$ has the property that $x \to \dim(\ker F(x))$ is continuous, then these kernels can be seen to fit together to form the total space of a bundle over X which we denote by $\ker F$. Likewise there is a bundle $\ker F^*$, and we can define index F to be the element of $K(X)$ determined by the pair $([\ker F], [\ker F^*])$. In general, the kernel dimension need not be continuous (e.g., on $[0, 1]$ let $F(t) = (I - P) + tP$, where P is a fixed finite-rank projection). However, it can be shown that every such F can be deformed so that the kernel dimension becomes continuous.

10.3 THE K-FUNCTOR

The K-functor constructed above can be "extended" to arbitrary rings R, in the sense that $K(R) = K(X)$ whenever $R = C(X)$. **Algebraic *K*-theory**, as developed by Milnor, Bass, Quillen, and others, is based on this extended K-functor. In this generality the theory presents severe difficulties, but, as mentioned previously, by restricting to C^*-algebras one can construct a theory with most of the pleasant features of topological K-theory. With this extension one of the most noteworthy features of K-theory becomes apparent. Namely the rings or C^*-algebras need not be commutative, and K-theory is seen to be a device for reducing to an abelian situation.

10.3.1

Recall that a section of a bundle $p : E \to X$ is a map $s : X \to E$ such that $p \circ s = \mathrm{id}_X$. The idea behind the extension of the K-functor to rings is that the **section functor** $E \to \Gamma(E)$, which assigns to a vector bundle its space of continuous sections, sets up a bijection between isomorphism classes of bundles over X and isomorphism classes of finitely generated projective modules over $C(X)$. This theorem, due to Swan [Swa62], results from the following facts:

i. Any vector bundle over a compact space X is a direct summand of a trivial bundle $X \times \mathbf{C}^n$.

ii. A module over a unital ring R is projective if and only if it is a direct summand of a free R-module.

Then, given a bundle E, if $E \oplus F = X \times \mathbf{C}^n$, it follows that $\Gamma(E)$ is a direct summand of

$$\Gamma(X \times \mathbf{C}^n) = C(X, \mathbf{C}^n) = C(X)^n$$

and thus is projective and finitely-generated. Conversely, given a finitely generated projective $C(X)$-module A, if $A \oplus B = C(X)^n$ and $e : C(X)^n \to A$ is the corresponding projection, then, because e is a module map, we can regard it as an element of

$$M_n(C(X)) = C(X, M_n(\mathbf{C}))$$

and thus obtain a vector bundle

$$E = \{v \in C(X, \mathbf{C}^n) : ev = v\}$$

10.3.2

For any unital ring R we therefore let $V(R)$ be the set of isomorphism classes of finitely generated projective (left) R-modules. With the operation induced by direct sum, this is an abelian semigroup with identity. As the reasoning in 10.3.1 indicates, the same semigroup can be built out of equivalence classes of **idempotents** ($e^2 = e$) in the matrix algebras $M_n(R)$. Here $M_n(R)$ is identified with the $n \times n$ corner $\left(\begin{smallmatrix} * & 0 \\ 0 & 0 \end{smallmatrix}\right)$ of $M_{n+k}(R)$, and in $M_\infty(R) = \bigcup M_n(R)$ the appropriate notion of equivalence of idempotents turns out to be an algebraic version of Murray–von Neumann equivalence: e and f are **algebraically equivalent** if there are $x, y \in M_\infty(R)$ with $xy = e$ and $yx = f$. One advantage of this picture is that it makes the functoriality of V evident (see below). It also allows a convenient description of the semigroup operation. Namely, to add the equivalence classes of idempotents e and f, choose representatives e' and f' with $e'f' = f'e' = 0$, and take the class of $e' + f'$. For example, if $e \in M_n(R)$ and $f \in M_k(R)$, take

$$e' = \begin{pmatrix} e & 0 \\ 0 & 0 \end{pmatrix} \quad \text{and} \quad f' = \begin{pmatrix} 0 & 0 \\ 0 & f \end{pmatrix}$$

in $M_{n+k}(R)$.

10.3.3

The **algebraic K-functor** is now defined, for *unital* rings, by composing the Grothendieck functor of 10.2.3 with the functor V just defined. In other words, $K(R)$ is the Grothendieck group of $V(R)$. To complete the definition of the functor, we also have to show that to any homomorphism $\varphi : R \to S$ of unital rings corresponds a homomorphism $\varphi_* : K(R) \to K(S)$ of groups, in such a way that

i. $(\mathrm{id}_A)_* = \mathrm{id}_{K(A)}$ for all unital rings R;
ii. $(\psi \circ \varphi)_* = \psi_* \circ \varphi_*$ for all unital homomorphisms $\varphi : R \to S$ and $\psi : S \to T$ of unital rings.

But the idempotent picture of the previous section makes this clear: A homomorphism $R \to S$ of rings applies entrywise to give a homomorphism $M_\infty(R) \to M_\infty(S)$ which therefore takes idempotents to idempotents, $V(R)$ to $V(S)$ and hence $K(R)$ to $K(S)$.

10.3.4

A technical difficulty now presents itself. For computational purposes one needs to know something about the effect of applying K to extensions

$$0 \to R \to S \to T \to 0$$

Since the kernel R is generally nonunital this requires that K be defined for nonunital rings. No matter how this is done, it is too much to expect that the induced sequence of abelian groups should always be exact. However, it is possible to get exactness at the middle term by defining K appropriately. The definition in terms of idempotents makes sense for nonunital rings but unfortunately does not have the desired property. Remarkably it is the case that the simplest possible choice for $K(R)$, namely as dictated by the unitization extension 1.3, does the job.

Definition

For any ring R, $K(R)$ is the kernel of the homomorphism $K(R^+) \to K(\mathbf{C})$ induced by the map $R^+ \to \mathbf{C}$ in the unitization extension for R.

Notice that $K(R^+) \to K(\mathbf{C})$ is a surjection, and hence one always has an exact sequence

$$0 \to K(R) \to K(R^+) \to K(\mathbf{C}) \to 0$$

When R is unital, the two definitions of $K(R)$ agree (as may easily be seen from the fact that then the unitization extension splits). The functoriality of the extended functor also needs to be checked. This amounts to the assertion that for any homomorphism $\varphi : R \to S$, the homomorphism $\varphi_*^+ : K(R^+) \to K(S^+)$ induced by the unital extension φ^+ carries the kernel of $K(R^+) \to K(\mathbf{C})$ into the kernel of $K(S^+) \to K(\mathbf{C})$.

10.3.5 Theorem (Half-exactness)

If the sequence

$$0 \to R \xrightarrow{i} S \xrightarrow{\pi} T \to 0$$

of rings is exact then so is the induced sequence

$$K(R) \xrightarrow{i_*} K(S) \xrightarrow{\pi_*} K(T)$$

of abelian groups.

The proof, for the case of Banach algebras, may be found in [Tay75] or [Bla86]. Apart from some subtle algebraic manipulations, the key point that comes up is the question of "lifting" invertibles: Does a surjection $S \to T$ of unital rings restrict to a surjection of their invertible groups? This is false in general. However it is true that

i. *A surjection of unital Banach algebras restricts to a surjection of the connected components of the identity of their invertible groups.*

For example, if $\|1 - u\| < 1$ in T, then with $x = 1 - u$ the Maclaurin series for $\log(1 - x)$ converges in T to an element a such that $e^a = u$; thus, if $b \in S$ maps to a, then e^b is invertible in S and maps to u. The general case is handled by showing that any element of the connected component of the identity is a product of exponentials.

This fact will be used in conjunction with

ii. *If $u \in GL_n(T)$, then $\left(\begin{smallmatrix} u & 0 \\ 0 & u^{-1} \end{smallmatrix}\right) \in GL_{2n}^0(T)$.*

Here T is a unital Banach algebra, $GL_n(T)$ is the group of invertible elements of $M_n(T)$, and $GL_n^0(T)$ is the connected component of the identity. In fact it is true that $x \oplus y$ and $xy \oplus 1$ are joined by a path of invertibles for any invertible $a, b \in M_n(T)$. The question of invertibility arises because of

iii. *If p and q are algebraically equivalent idempotents in $M_n(A)$, where A is a unital ring, then $\begin{pmatrix} p & 0 \\ 0 & 0 \end{pmatrix}$ and $\begin{pmatrix} q & 0 \\ 0 & 0 \end{pmatrix}$ are similar in $M_{2n}(A)$.*

For the proofs of these facts, the reader may consult [Bla86, §§3 and 4].

Proof of Theorem

We assume $R = \ker \pi$ and want to show $K(R) \supset \ker \pi_*$ (the other inclusion is straightforward). Consider any $x \in K(S)$ such that $\pi_*(x) = 0$. Tracing back through the definitions, we see that x is represented by a pair (e, f) of idempotents in some $M_n(S^+)$ such that $(c(e), c(f))$ and $(\pi(e), \pi(f))$ represent the zero elements of $K(\mathbf{C})$ and $K(T)$ respectively, where $c : S^+ \to \mathbf{C}$ is the map in the unitization sequence. For simplicity we assume that $f = 1_k$ (the $k \times k$ identity matrix) for some $k < n$ and that $\pi(e)$ and $\pi(f)$ are algebraically equivalent. (In fact it is not difficult to see that such a representative of x always exists.) That $(c(e), c(f))$ is zero in $K(\mathbf{C})$ means simply that $\operatorname{rank} c(e) = \operatorname{rank} c(f)$. By the facts stated above we have $u^{-1}\pi(e)u = \pi(f)$ for some $u = \pi(v)$ with $v \in GL_m(S^+)$ and $m = 4n$. Thus $v^{-1}ev - f \in M_m(R)$, and therefore, since $f = 1_k$, both $v^{-1}ev$ and f are in $M_m(R^+)$. Since $c(v^{-1}ev)$ and $c(f)$ have the same rank, the pair $(v^{-1}ev, f)$ determines an element of $K(R)$. But because $v \in M_m(S^+)$, this pair also represents x, so $x \in K(R)$.

10.3.6

Here are some examples of K-groups.

Example 1. When R is unital $K(R)$ contains a canonical copy of $\mathbf{Z}$. In fact the map that assigns to $n \in \mathbf{Z}$ the element of $K(R)$ determined by the pair $(1_n, 0)$ if $n \geq 0$ or $(0, 1_{-n})$ if $n \leq 0$ is readily seen to be an injective homomorphism $\mathbf{Z} \to K(R)$. If $R = \mathbf{C}$, it is an isomorphism and therefore $K(\mathbf{C}) \cong \mathbf{Z}$.

Example 2. $K(M_n(\mathbf{C})) \cong \mathbf{Z}$. In fact $V(M_n(\mathbf{C})) = V(\mathbf{C})$ because both are built out of idempotents in matrix algebras over $\mathbf{C}$. A closely related fact that is $K(\mathcal{K}(\mathcal{H})) \cong \mathbf{Z}$. This can be seen from the fact that K is **continuous**, in the sense that it preserves direct limits (see below), since $\mathcal{K}(\mathcal{H}) = \lim_{\to} M_n(\mathbf{C})$. It can also be done with "bare hands" by showing that for any idempotent

$p \in M_\infty(\mathcal{K}(\mathcal{H})^+)$, where $\mathcal{K}(\mathcal{H})^+$ is identified with $\mathcal{K}(\mathcal{H}) + \mathbf{C}I_{\mathcal{H}}$, there is an equivalent idempotent p' and an integer $k \geq 0$ such that p' is comparable to $I_k \in M_k(\mathcal{K}(\mathcal{H})^+)$ and $p' - I_k$ has finite rank. Then $p \to (k, \operatorname{tr}(p' - I_k))$ induces an isomorphism $K(\mathcal{K}(\mathcal{H})^+) \cong \mathbf{Z} \oplus \mathbf{Z}$. But $c : \mathcal{K}(\mathcal{H})^+ \to \mathbf{C}$ splits, so c_* has image $\mathbf{Z}$ and

$$K(\mathcal{K}(\mathcal{H})) = \ker c_* \cong \frac{(\mathbf{Z} \oplus \mathbf{Z})}{\operatorname{im} c_*} \cong \mathbf{Z}$$

Example 3. $\mathcal{K}(\mathcal{B}(\mathcal{H})) = 0$ because for any two idempotents e, f in $M_n(\mathcal{B}(\mathcal{H}))$, $e \oplus I \oplus 0$ and $f \oplus I \oplus 0$ are equivalent in $M_{n+2}(\mathcal{B}(\mathcal{H}))$. In particular, the map induced by the inclusion $\mathcal{K}(\mathcal{H}) \to \mathcal{B}(\mathcal{H})$ is not injective, showing that i_* in 10.3.5 need not be injective.

Example 4. If X is a contractible compact Hausdorff space, then because (as we will see below) K is homotopy invariant,

$$K(C(X)) \cong K(C(pt)) = K(\mathbf{C}) \cong \mathbf{Z}$$

In particular, since K respects direct sums, the surjection $C[0, 1] \to \mathbf{C} \oplus \mathbf{C}$ obtained by evaluation at 0 and 1 does not induce a surjection. Hence π_* in 10.3.5 need not be surjective.

Example 5. The isomophism $[X, \mathcal{F}] \to K(C(X))$ of 10.2.5 includes the previous example. There are still more general index isomorphisms

$$[X, \mathcal{F}_A] \to K(C(X) \otimes A)$$

where A is any unital C^*-algebra and $\mathcal{F}_A$ is the set of Fredholm operators on the Hilbert C^*-module $\mathcal{H}_A$ of 3.14.4 [Min87].

10.4 OPERATOR ALGEBRA K-THEORY

As mentioned previously the K-functor is the starting point of both algebraic and topological K-theory. By itself it is difficult to compute and consequently is not particularly useful. To remedy this the idea is to proceed as in the ordinary homology and cohomology theories of algebraic topology and make K the first in a sequence $K_0, K_1, \ldots$ of functors that complement each other nicely. This can be done—with difficulty—in the purely algebraic case; Higson's thesis [Hig88] contains as interesting account and further references. In the case of Banach algebras, the additional functors are readily defined with the help of a concept borrowed from topology.

10.4.1

The **suspension** of a Banach algebra A is $SA = C_0(\mathbf{R}, A)$. With the pointwise operations and supremum norm, this is again a Banach algebra. When $A = C(X)$, it can be identified with the functions on $X \times I$ vanishing on $X \times \{0, 1\}$, and collapsing the latter set to a point in $X \times I$ yields the usual suspension of X as a topological space.

10.4.2 Definition

For Banach algebras A the higher K-functors are defined by iterated suspension:

$$K_n(A) = K(S^n A) \qquad \textit{for } n \geq 0$$

where $S^0 A = A$, $S^1 A = SA$, $S^2 A = S(SA)$, ...

(To conform to usage in algebraic topology, one would use this equation to define K_{-n}, instead of K_n, but, in view of the next theorem, they are the same in the present context.)

10.4.3 Theorem (Bott Periodicity)

$K(S^2 A) \cong K(A)$ *for all Banach algebras A. More precisely, the functors* K_n *and* K_{n+2} *are naturally isomorphic for all* $n \geq 0$.

This is the fundamental result of the theory; see 10.5 for a discussion of the proof.

10.4.4

By the Bott periodicity theorem our principal concern is with the functors K_0 $(= K)$ and K_1 $(= KS)$. The groups $K_0(A)$ and $K_1(A)$ can be realized in a number of ways. It was stated in 10.3.5(iii) that on $M_\infty(A)$ similarity is the same as algebraic equivalence. The same is true of homotopy, idempotents being called **homotopic** if they are connected by a norm-continuous path of idempotents. When A is a C^*-algebra, attention can be restricted to projections, since in this case every idempotent is similar to a projection [Bla86].

Turning to K_1, for unital Banach algebras A identify $a \in GL_n(A)$ with $a \oplus 1 \in GL_{n+1}(A)$, and let $GL_\infty(A) = \cup GL_n(A)$. With the inductive limit topology (U is open in GL_∞ if $U \cap GL_n$ is open in GL_n for all n), this is a topological group. Then the functors GL_∞/GL_∞^0 (= the group of connected components) and K_1 are naturally isomorphic. To get an idea of why this should be so, consider $u \in GL_n(A)$. Then $u \oplus u^{-1} \in GL_{2n}^0(A)$ by 10.3.5(ii); let v_t be a path connecting it to 1_{2n}, and let $p_t = v_t(1_n \oplus 0)v_t^{-1}$. Identify (0, 1) with $\mathbf{R}$ so that p_t determines an idempotent $p \in M_{2n}(SA)^+$. If $1_n \oplus 0 \in M_{2n}(SA)^+$ is the constant path, then the pair $(p, 1_n \oplus 0)$ determines an element of $K(SA)$, and $u \to (p, 1_n \oplus 0)$ induces the isomorphism in question.

Among the consequences of this realization of K_1 are the facts that it is half-exact (10.3.5) and that (cf. 10.2.5)

$$K_1(C(X)) \cong [X, GL_\infty(\mathbf{C})] \cong [X, \mathcal{U}_c]$$

where $\mathcal{U}_c = \mathcal{U}(\mathcal{H}) \cap (I_{\mathcal{H}} + \mathcal{K}(\mathcal{H}))$ and $\mathcal{U}(\mathcal{H})$ is the unitary group of the separable Hilbert space $\mathcal{H}$.

10.4.5

Here are some examples of K_1-groups.

EXAMPLE 1. $K_1(\mathbf{C}) = 0$ because $GL_n(\mathbf{C})$ is connected. In fact $GL(\mathcal{M})$, the group of invertible elements of $\mathcal{M}$, is connected for any von Neumann algebra $\mathcal{M}$ (via polar decomposition and the fact that any normal operator $N \in \mathcal{M}$ can be displayed in the form $N = e^M$ for some $M \in \mathcal{M}$), and therefore $K_1(\mathcal{M}) = 0$. In particular, $K_1(\mathcal{B}(\mathcal{H})) = 0$.

EXAMPLE 2. $K_1(\mathcal{K}(\mathcal{H})) = 0$ by the same sort of reasoning. In general, $K_1(A) = K_1(A^+)$, and one can deform $(K_{ij} + \lambda_{ij} I_{\mathcal{H}}) \in GL_n(\mathcal{K}(\mathcal{H}) + \mathbf{C}I_{\mathcal{H}})$ in GL_n, first to $(\lambda_{ij} I_{\mathcal{H}})$, because of the countability of the spectrum of a compact operator, and then to I.

EXAMPLE 3. The Calkin algebra $\mathcal{C} = \mathcal{B}(\mathcal{H})/\mathcal{K}(\mathcal{H})$ has $K_1(\mathcal{C}) \cong \mathbf{Z}$. This follows from the fact that the connected components of $GL(\mathcal{C})$ are precisely the images of the sets $\mathcal{F}_n \subset \mathcal{B}(\mathcal{H})$ of Fredholm operators of index n. Thus the isomorphism is given by the index map.

EXAMPLE 4. The K-groups of $C(S^1)$ are both $\mathbf{Z}$. This can be seen from first principles, by showing that any projection-valued map $S^1 \to M_n(\mathbf{C})$ is homotopic to a constant one and that any unitary-valued map is homotopic to one of the form $e^{i\theta} \to e^{ik\theta} \oplus 1_{n-1}$. Thus the K-groups are determined by rank and winding number respectively. Alternatively, the K_0-group can be obtained by observing that the unitization of $C_0(\mathbf{R}) = S\mathbf{C}$ is $*$-isomorphic to $C(S^1)$, so by 10.3.4 we have an exact sequence

$$0 \to K_0(S\mathbf{C}) \to K_0(C(S^1)) \to K_0(\mathbf{C}) \to 0$$

But by definition the first of these groups is $K_1(\mathbf{C})$, which has just been shown to be trivial. Hence $K_0(C(S^1)) \cong K_0(\mathbf{C}) \cong \mathbf{Z}$. We will shortly indicate a similar argument for K_1.

10.4.6 Theorem (Homotopy Invariance)

If A and B are Banach algebras and

$$\varphi : A \to C([0, 1], B)$$

is a continuous homomorphism, then the induced homomorphisms

$$\varphi_{t*} : K_*(A) \to K_*(B)$$

are all equal. Here $\varphi_t = \epsilon_t \circ \varphi$ *and* $\epsilon_t : C([0, 1], B) \to B$ *is evaluation at* $t \in [0, 1]$.

The use of the notation K_* here means that the theorem holds for both K_0 and K_1. For K_1 the result is easy to see, at least in the unital case, because for any $a \in GL_n(A)$ the path $t \to \varphi_t(a)$ is norm-continuous and so lies in a single connected component of $GL_n(B)$. For K_0 it follows from the fact that projections e, f in a C^*-algebra are similar if $\|e - f\| < 1$, since φ sends any projection $p \in M_n(A)$ to a norm-continuous path $t \to \varphi_t(p)$ of projections.

10.4.7

Thus the K-groups are insensitive to continuous deformations. For example, let X be a contractible space. Thus there are $x_0 \in X$ and $F : [0, 1] \times X \to X$ such that $F(0, x) = x_0$ and $F(1, x) = x$ for all $x \in X$. Then $\varphi(f) = f \circ F$ defines a continuous homomorphism

$$\varphi : C(X) \to C([0, 1] \times X) = C([0, 1], C(X))$$

and $\mathrm{id}_* = \varphi_{1*} = \varphi_{0*}$ by the theorem, so $K_*(C(X)) = K_*(\mathbf{C})$ (i.e., $\mathbf{Z}$ or 0 according as $* = 0$ or 1).

10.4.8

As remarked in 10.3.4, the K-functor does not preserve short exact sequences. The same is true of K_1, and it is a remarkable fact that each of the induced sequences "exactly" repairs the deficiencies in the other:

Theorem (Long Exact Sequence)

For any short exact sequence

$$0 \to A \xrightarrow{i} B \xrightarrow{\pi} C \to 0$$

of Banach algebras there is a natural periodic exact sequence

$$\begin{array}{ccccc} K_0(A) & \xrightarrow{i_*} & K_0(B) & \xrightarrow{\pi_*} & K_0(C) \\ \partial \uparrow & & & & \downarrow \partial \\ K_1(C) & \xleftarrow[\pi_*]{} & K_1(B) & \xleftarrow[i_*]{} & K_1(A) \end{array}$$

Outline of the Proof

The first step is to make use of the half-exactness and homotopy invariance of K to borrow from topology the construction of a "connecting map" ∂ such that the sequence

$$K_1(A) \to K_1(B) \to K_1(C) \xrightarrow{\partial} K_0(A) \to K_0(B) \to K_0(C)$$

is exact. The loop is then closed using the Bott isomorphism:

$$K_0(C) \cong K_2(C) = K_1(SC) \xrightarrow{\partial} K_0(SA) = K_1(A)$$

In the case of C^*-algebras, the connecting map has a very natural interpretation in terms of index. In the special case

$$0 \to \mathcal{K}(\mathcal{H}) \to \mathcal{B}(\mathcal{H}) \to \mathcal{C} \to 0$$

we have

$$K_1(\mathcal{C}) \xrightarrow{\text{ind}} \mathbf{Z} \cong K_0(\mathcal{K}(\mathcal{H}))$$

as seen in the examples. In the general case, let $x \in GL_n(C^+)$, and let $x = |x|u$ be its polar decomposition so that u is unitary and determines the same element of $K_1(C)$ that x does. Since π is surjective, there is $b \in M_n(B^+)$ with $\pi(b) = u$. It can be shown that because $\|u\| = 1$, we can choose b with $\|b\| = 1$; hence $(1 - b^*b)^{1/2}$ exists by 2.3.1 and 2.3.2. Then

$$v = \begin{pmatrix} b & 0 \\ (1 - b^*b)^{1/2} & 0 \end{pmatrix}$$

is a partial isometry and its defect projections $1 - v^*v$ and $1 - vv^*$ are in $M_{2n}(A^+)$. Here we have assumed, as we can, that $A = \ker \pi$. Moreover we have $c(1 - v^*v) = c(1 - vv^*)$ in $M_{2n}(\mathbf{C})$. To see this note that if $b = (y, \lambda)$ with $y \in M_n(B)$, then $|\lambda| = 1$ because $\pi(b)$ is unitary. It follows that the defect projections determine an element of $K_0(A)$. The connecting map is then induced by the correspondence $x \to (1 - v^*v, 1 - vv^*)$.

As an illustration of how this result can be used in computations, apply it to the unitization extension

$$0 \to A \to A^+ \to \mathbf{C} \to 0$$

As mentioned in 10.3.3, the induced K_0 sequence is exact, and hence the boundary map $K_0(\mathbf{C}) \to K_1(A)$ is zero. Since also $K_1(\mathbf{C}) = 0$, we conclude that $K_1(A) \cong K_1(A^+)$. In particular, taking $A = C_0(\mathbf{R}) = S\mathbf{C}$ and identifying the unitization with $C(S^1)$, we find that $K_1(C(S^1)) \cong K_1(S\mathbf{C}) \cong \mathbf{Z}$ since $K_1(S\mathbf{C}) = K_0(S^2\mathbf{C}) \cong K_0(\mathbf{C})$ by periodicity.

10.4.9 Theorem (Stability)

Let A be a C^*-algebra and define $i : A \to A \otimes \mathcal{K}(\mathcal{H})$ by $i(a) = a \otimes e$, where e is a fixed projection of rank one and $\mathcal{H}$ is a separable infinite dimensional Hilbert space. Then the induced map i_* is an isomorphism.

We have already observed this for $A = \mathbf{C}$: $K_0(\mathcal{K}(\mathcal{H})) = K_0(\mathbf{C}) = \mathbf{Z}$ and $K_1(\mathcal{K}(\mathcal{H})) = K_1(\mathbf{C}) = 0$. For the general case consider the direct sequence

$$A = M_1(A) \to M_2(A) \to \cdots$$

with maps $a \to a \oplus 0$, and the induced sequences

$$K_*(A) = K_*(M_1(A)) \to K_*(M_2(A)) \to \cdots$$

It follows from the definitions that the induced maps are isomorphisms, and so the canonical maps

$$K_*(M_n(A)) \to \lim_{\to} K_*(M_n(A))$$

are isomorphisms. Now the result follows from the fact that K-theory respects limits of direct sequences (and that $\lim_{\to} M_n(A) \cong A \otimes \mathcal{K}(\mathcal{H})$):

10.4.10 Theorem (Continuity)

Let $A_1 \to A_2 \to \cdots$ be a direct sequence of C^*-algebras and $*$-homomorphisms, let $A = \lim_{\to} A_n$, and let $K_*(A_1) \to K_*(A_2) \to \cdots$ be the induced direct sequence of abelian groups and homomorphisms. Then the homomorphism

$$\lim_{\to} K_*(A_n) \to K_*(A)$$

determined by the universal property of direct limits from the maps induced by the canonical maps $A_n \to A$ is an isomorphism.

The idea of the proof of surjectivity is that elements of $GL_k(A^+)$ can be approximated by elements of $GL_k(A_n^+)$ and projections in $M_n(A^+)$ by projections in $M_k(A_n^+)$. As applications of this property we have $K_1 = 0$ for all AF algebras and $K_0(U(2^\infty)) = \mathbf{Z}[\frac{1}{2}]$, the dyadic rationals, where $U(2^\infty)$ is the UHF algebra of 3.10.1. The corresponding Bunce-Deddens algebra (3.10.3) has the same K_0 but $K_1 = \mathbf{Z}$.

10.5 AXIOMATIZATION

The preceding discussion completes our presentation of the basic definitions, examples and properties of K-theory. In this section we outline Cuntz's axiomatization of Bott periodicity [Cun84].

10.5.1 Theorem (Cuntz)

Any half-exact, homotopy invariant, and stable functor from C^*-algebras to abelian groups satisfies Bott periodicity.

Idea of the Proof (see [Hig88])

Start with the Toeplitz sequence (3.11.1),

$$0 \to \mathcal{K} \to \mathcal{T} \to C(\partial\mathbf{D}) \to 0$$

where $\mathcal{T} = C^*(S)$, S the unilateral shift. Fix a point (e.g., -1) in $\partial\mathbf{D}$, and let $\mathcal{T}_0 \subset \mathcal{T}$ be the ideal of elements with symbol (3.11.3) vanishing there. Then we have an exact sequence

$$0 \to \mathcal{K} \to \mathcal{T}_0 \to C_0(\mathbf{R}) \to 0$$

Now any half-exact homotopy invariant functor F associates to any short exact sequence

$$0 \to A \xrightarrow{i} B \xrightarrow{\pi} C \to 0$$

a natural long exact sequence

$$\cdots \to F_n(A) \xrightarrow{i_*} F_n(B) \xrightarrow{\pi_*} F_n(C) \xrightarrow{\partial}$$

$$\xrightarrow{\partial} F_{n-1}(A) \xrightarrow{i_*} F_{n-1}(B) \xrightarrow{\pi_*} F_{n-1}(C) \to \cdots$$

where $F_n = F_{n-1}S$, $F_0 = F$, and S is suspension (10.4.1). The construction, via mapping cones, is standard in topology [EiS52]. Now

$$0 \to \mathcal{K} \otimes A \to \mathcal{T}_0 \otimes A \to C_0(\mathbf{R}) \otimes A \to 0$$

is exact (7.2.13 and 7.4.1); hence so is

$$F_1(\mathcal{T}_0 \otimes A) \to F_1(C_0(\mathbf{R}) \otimes A) \xrightarrow{\partial} F_0(\mathcal{K} \otimes A) \to F_0(\mathcal{T}_0 \otimes A)$$

Since $S(C_0(\mathbf{R}) \otimes A) = S^2 A$ and F has been assumed to be stable, it will be enough to show that $F_n(\mathcal{T}_0 \otimes A) = 0$ for $n = 0, 1$. The essential features of the proof of this appear already when $A = \mathbf{C}$, the general case then being handled by "tensoring on" A appropriately. The relevant property of $\mathcal{T}_0$ is that for any extension of the form

$$0 \to A \otimes \mathcal{K} \to B \to \mathcal{T}_0 \to 0$$

the induced long exact sequence breaks up into short exact sequences.

$$0 \to F_n(A \otimes \mathcal{K}) \to F_n(B) \to F_n(\mathcal{T}_0) \to 0$$

Consider then the commuting diagram

$$\begin{array}{ccccccccc}
0 & \to & C_0(\mathbf{R}) \otimes \mathcal{T}_0 \otimes \mathcal{K} & \to & B & \to & \mathcal{T}_0 & \to & 0 \\
 & & \| & & \downarrow & & \downarrow i & & \\
0 & \to & C_0(\mathbf{R}) \otimes \mathcal{T}_0 \otimes \mathcal{K} & \to & C_0(0,1] \otimes \mathcal{T}_0 \otimes \mathcal{K} & \xrightarrow[\pi]{} & \mathcal{T}_0 \otimes \mathcal{K} & \to & 0
\end{array}$$

of exact sequences, where i is the map of 10.4.9, π is evaluation at 1, and B is the pullback. By the above, $F_n(B) \to F_n(\mathcal{T}_0)$ is a surjection and (because i_*

is an isomorphism), it follows that

$$F_n(C_0(0,1]\otimes\mathcal{T}_0) = F_n(C_0(0,1]\otimes\mathcal{T}_0\otimes\mathcal{K}) \to F_n(\mathcal{T}_0\otimes\mathcal{K}) = F_n(\mathcal{T}_0)$$

is a surjection too. But $C_0(0,1]\otimes\mathcal{T}_0$ is contractible, in the sense that there exists a $*$-homomorphism

$$\varphi : C_0(0,1]\otimes\mathcal{T}_0 \to C([0,1], C_0(0,1]\otimes\mathcal{T}_0)$$

such that $\varphi_0 = 0$ and $\varphi_1 = \text{id}$. If $f \in C_0(0,1]\otimes\mathcal{T}_0 = C_0((0,1],\mathcal{T}_0)$, define $(\varphi_t(f))(s) = f(st)$, where $f(0)$ is taken to be 0. Hence, by homotopy invariance (10.4.6), we have $F_n(C_0(0,1]\otimes\mathcal{T}_0) = 0$, and therefore $F_n(\mathcal{T}_0) = 0$.

10.5.2

Cuntz's theorem opens the way to possible characterizations of K-theory. It may be that K is the unique functor on C^*-algebras to abelian groups that is half-exact, homotopy invariant, stable, continuous, and satisfies the **dimension axiom**

$$K(\mathbf{C}) = \mathbf{Z} \quad \text{and} \quad K(S\mathbf{C}) = 0$$

This is known to be true on the smallest category of separable C^*-algebras that contains the commutative algebras and is closed under the standard constructions of C^*-algebra theory (a category that would appear to contain most interesting C^*-algebras). For more details, consult [Bla86] and the references given there.

10.6 FURTHER DEVELOPMENTS

We conclude with a brief discussion of several other aspects of operator K-theory, including a few of its more notable achievements.

10.6.1 Structure of Crossed Products

Generally speaking, it is difficult to get information about the structure of these algebras. However, our understanding of crossed products by 1-parameter groups has been substantially advanced by several important results.

The Pimsner-Voiculescu Exact Sequence [PiV80]

For any C^*-algebra A and any automorphism α of A, there is a cyclic exact sequence

$$\begin{array}{ccccc} K_0(A) & \xrightarrow{\text{id}-\alpha_*} & K_0(A) & \xrightarrow{i_*} & K_0(A \rtimes_\alpha \mathbf{Z}) \\ \uparrow & & & & \downarrow \\ K_1(A \rtimes_\alpha \mathbf{Z}) & \xleftarrow[i_*]{} & K_1(A) & \xleftarrow[\text{id}-\alpha_*]{} & K_1(A) \end{array}$$

Several proofs are available (see [Bla86] for a discussion), in all of which the strategy is to devise an extension

$$0 \to A \otimes \mathcal{K} \to B \to A \rtimes_\alpha \mathbf{Z} \to 0$$

so that the long exact sequence 10.4.8 becomes the desired sequence. A similar result for $A \rtimes_\alpha S^1$ can be deduced from this one by Takai duality 8.6.7 (where S^1 is the circle group, the dual of $\mathbf{Z}$), but it is not nearly as useful, since this time the crossed product appears four times in the sequence.

As an example of how this sequence works, we look at the rotation algebras $A_\theta = C(S^1) \rtimes_\alpha \mathbf{Z}$, where α is the automorphism of $C(S^1)$ defined by rotation through the angle $2\pi\theta$. As mentioned earlier (10.4.5), the K_0-classes of $C(S^1)$ are determined by rank and the K_1-classes by winding number. Since rotation does not change these, the α_* are identity maps, and the Pimsner-Voiculescu (PV) sequence then gives $K_*(A_\theta) \cong K_0(C(S^1)) \oplus K_1(C(S^1)) \cong \mathbf{Z}^2$. Note that this is independent of θ, and so these algebras are not distinguished by their K-groups.

The PV sequence can also be used to show that $K_0(\mathcal{O}_n) \cong \mathbf{Z}_{n-1}$ and $K_1(\mathcal{O}_n) = 0$, where $\mathcal{O}_n$ is the Cuntz algebra (3.11.4). To do this, one constructs a copy of the UHF algebra $U(n^\infty)$ inside $\mathcal{O}_n$ and an automorphism α of $U(n^\infty) \otimes \mathcal{K}$ such that the crossed product is isomorphic to $\mathcal{O}_n \otimes \mathcal{K}$. We refer the reader to [Bla86] or [Dav95] for the details of these examples.

The Connes-Thom Isomorphism [Con81]

For any C^*-dynamical system $(\mathbf{R}, \alpha, A)$, there are isomorphisms

$$K_{n+1}(A \rtimes_\alpha \mathbf{R}) \cong K_n(A)$$

In particular, the K-theory of the crossed product is independent of the action. When the action is trivial, one has

$$A \rtimes_\alpha \mathbf{R} \cong A \otimes C_0(\mathbf{R}) \cong C_0(\mathbf{R}, A) = SA$$

by 8.3.2 and 10.4.1, and the theorem reduces to the Bott periodicity theorem. The strategy of the proof is the same as above (with $\mathbf{Z}$ replaced by $\mathbf{R}$), except that homotopy invariance and the contractibility of the cone $CA = C_0((0, 1], A)$ now imply $K_*(B) = 0$, so the periodic sequence degenerates into the two desired isomorphisms.

This result gets its name by analogy with the Thom isomorphism, which says that the K-theory of a vector bundle is the same as that of the trivial bundle of the same dimension (the latter is, up to a dimension shift, the K-theory of the base space).

10.6.2 Classification of AF Algebras and the Elliott Program

In favorable cases, taking the image of $V(A)$ in $K_0(A)$ as the positive cone $K_0(A)_+$ gives $K_0(A)$ the structure of an ordered group (see [Bla86, §6]). Keeping track of this extra structure allows one to retain at least some of the information that is lost in the passage from $V(A)$ to its Grothendieck group (but note that even as an ordered group K_0 does not distinguish between A and $A \otimes \mathcal{K}$).

For AF algebras (3.10.4) K_0 is an ordered group and Elliott's classification theorem [Ell76] asserts that it determines the **stable isomorphism** class of the algebra. In other words, for AF algebras A and B, $K_0(A) \cong K_0(B)$ as ordered groups if and only if $A \otimes \mathcal{K} \cong B \otimes \mathcal{K}$. To distinguish among elements of a given stable isomorphism class, it is necessary to record which K_0-classes come from elements of the algebra itself. Denoting by $[P(A)]$ the set of K_0-classes of projections in A, if the above order isomorphism takes $[P(A)]$ to $[P(B)]$, then one can even conclude that $A \cong B$. When A and B are unital, this will be the case if the K_0-class of 1_A is taken to that of 1_B. Notice that the class of the identity determines the positive cone too so that in the unital case one only need keep track of it. The ordered abelian groups which occur as K_0-groups of AF algebras are called **dimension groups**. They have been characterized by Effros, Handleman and Shen. Good accounts of this material may be found in [Eff81], [Bla86] and [Dav95]; [Mur90] treats Elliott's theorem but does not include dimension groups.

Elliott's result makes no mention of K_1, and in fact continuity implies that $K_1(A) = 0$ for any AF algebra A. This raises the possibility of using both K_0 and K_1 to extend the classification to a larger class of C^*-algebras.

But even this would not be enough to include many commonly occurring C^*-algebras. A case in point is the class of irrational rotation algebras A_θ, which for $0 < \theta < \frac{1}{2}$ are pairwise nonisomorphic and yet have the same K-groups (10.6.1). In such situations a useful additional isomorphism invariant is the **tracial state space** $T(A)$, which consists of states $\tau \in A^*$ having the trace property $\tau(ab) = \tau(ba)$ for all $a, b \in A$. It is weak*-compact and convex (in fact a Choquet simplex if A is separable) and is determined up to affine homeomorphism by the isomorphism class of A. The tracial state space is not completely independent of the K-theory data. In fact there is a more-or-less evident pairing

$$T(A) \times K_0(A) \longrightarrow \mathbf{R}$$

since any trace is the same on any two equivalent projections. (Any $\tau \in T(A)$ extends to a compatible family of traces τ_n on $M_n(A)$ by setting $\tau_n(a) = \sum_i \tau(a_{ii})$.) Now it can be shown that for θ irrational $T(A_\theta)$ is a singleton (obtained by following the conditional expectation $E : A_\theta \to C(S^1)$ of 3.9 by integration with respect to Lebesgue measure) and that the order-preserving homomorphism $\tau : K_0(A_\theta) \to \mathbf{R}$ coming from the pairing has range $\mathbf{Z} + \mathbf{Z}\theta$ ([Rie81]; see also [Bla86] and [Dav95]). From this it follows easily that these algebras are pairwise nonisomorphic as asserted.

Recent work of Elliott and others (see [Ell95]) has focused attention on direct limits of finite direct sums of algebras of the form $M_n \otimes C(X)$, especially when the spaces X are of bounded finite dimension and the limit algebra is simple. Rather definitive results have been obtained for the case of **circle algebras**, obtained by taking all $X = S^1$. It has recently been shown [ElE93] that the irrational rotation algebras are circle algebras; this raises the problem of finding an intrinsic characterization of the class of circle algebras.

10.6.3 Purely Infinite Algebras

A C^*-algebra A is said to be **purely infinite** if every hereditary subalgebra of A contains an infinite projection. In more detail, for separable algebras this is equivalent to the requirement that every subalgebra of the form $(aAa)^-$ contains a projection that is Murray–von Neumann equivalent to a proper subprojection. In particular, every nonzero projection in such an algebra is infinite. The first examples to be studied were the Cuntz algebras $\mathcal{O}_n$ (3.11.4). The Cuntz algebras are simple, and Cuntz later extended his study to arbitrary simple purely infinite algebras [Cun81]. He showed, for example, that every projection in such an algebra contains two (hence n) orthogonal copies of itself and that every K_0-class is represented by a projection in the algebra. Thus the order structure described above is not useful here. Furthermore there are no nonzero traces and so the isomorphism invariants discussed in the last section reduce to the groups K_0 and K_1. Notice that the computation of the K-groups of the Cuntz algebras in 10.6.1 shows that both are $\{0\}$ for $\mathcal{O}_2$.

In a recent breakthrough Rørdam was able to describe a class of simple, purely infinite and stable algebras for which the K-groups form a complete invariant. In other words, two algebras in the class are $*$-isomorphic if and only if they have isomorphic K-groups. Furthermore he showed how to "exhaust the invariant": For any pair of countable abelian groups G_i, there is an algebra A (in fact a circle algebra) and an automorphism α of A such that the crossed product $A \rtimes_\alpha \mathbf{Z}$ is in the specified class and has the G_i as its K-groups [Rør95]. There is an obvious parallel between this result and the Connes-Takesaki theorem 9.2.4 describing type III factors as crossed products; it suggests that it may be possible to obtain a classification result for amenable C^*-algebras analogous to the classification theorem for hyperfinite factors.

The difficulty with Rørdam's result is that it does not provide a usable way of determining whether or not a given algebra is in the class in question. Very recently Kirchberg and Phillips have come close to settling this question completely [Kir94; Kir95; Phi94]. Thus it now seems reasonable to conjecture that Rørdam's class consists of *all* algebras that are separable, nuclear, simple, purely infinite, and stable. One of the last remaining difficulties in establishing this is to show that $\mathcal{O}_2 \otimes \mathcal{K}$ is the only algebra in the class with trivial K-groups.

10.6.4 BDF Theory

The study of C^*-algebra extensions of the form

$$0 \to \mathcal{K} \to \mathcal{E} \to C(X) \to 0$$

where $\mathcal{E}$ is a C^*-subalgebra of $\mathcal{B}(\mathcal{H})$ containing the compact operators $\mathcal{K} = \mathcal{K}(\mathcal{H})$, and X is a compact metrizable space (i.e. $C(X)$ is a separable unital commutative C^*-algebra), was undertaken by Brown, Douglas and Fillmore in [BDF73] and [BDF77]. As noted in 3.12.6, such extensions correspond to injective (because $\mathcal{K}$ is an essential ideal) unital morphisms τ from $C(X)$ to the Calkin algebra $\mathcal{C}(\mathcal{H}) = \mathcal{B}(\mathcal{H})/\mathcal{K}(\mathcal{H})$. Two such extensions are **equivalent**, $\tau_1 \sim \tau_2$, if there is a unitary operator U on $\mathcal{H}$ such that $\tau_2 = \operatorname{Ad}\pi(U) \circ \tau_1$, where $\pi : \mathcal{B}(\mathcal{H}) \to \mathcal{C}(\mathcal{H})$ is the quotient map. Addition of extensions is by direct sum, as described in 3.11.5 in the case of the circle, and the first main result of the theory is that this operation gives the set $\operatorname{Ext}(X)$ of equivalence classes the structure of an abelian group. From here the theory is developed by showing that the functor $X \to \operatorname{Ext}(X)$ can be used to construct a homology theory for compact metrizable spaces, in much the same way that the K-functor is used to construct K-theory.

There is a bilinear pairing

$$\mathrm{Ext}(X) \times \pi^1(X) \to \mathbf{Z}$$

where $\pi^1(X)$ is the group of homotopy classes of invertible elements of $C(X)$, induced by taking (τ, f) to the Fredholm index of $\tau(f)$. Notice here that $\tau(f)$ is an invertible element of $\mathcal{C}(\mathcal{H})$ because f is invertible and τ is unital. This yields a homomorphism

$$\mathrm{Ext}(X) \to \mathrm{Hom}(\pi^1(X), \mathbf{Z})$$

which in general is not an isomorphism. Thus the group of homomorphisms on the right never has torsion, but there are subsets X of $\mathbf{C}^2$ such that Ext(X) does have torsion.

However, for subsets of the plane, this map can be shown to be an isomorphism. Moreover, $\pi^1(X)$ can be identified as the free abelian group generated by the classes $[z - \lambda]$, one for each bounded component B of the complement of X in $\mathbf{C}$, where λ is any point of B. Thus Ext(X) is the group of all integer assignments to these generators.

This result leads to an interesting theorem in operator theory. Observe first that requiring X to be a subset of the plane is essentially equivalent to requiring that $C(X)$ be singly-generated (by the function z). It follows that every extension of $C(X)$ by $\mathcal{K}$ can be obtained by taking $\mathcal{E} = C^*(T)$, where $T \in \mathcal{B}(\mathcal{H})$ is **essentially normal**, in the sense that $T^*T - TT^* \in \mathcal{K}(\mathcal{H})$, and $\pi(T)$ has spectrum X. The morphism $\mathcal{E} \to C(X)$ is determined by sending T to z. Consequently Ext(X) is isomorphic to the set of all such operators T, modulo unitary equivalence up to a compact perturbation. Combining this with the identification of Ext(X) in the preceding paragraph gives

Theorem [BDF73, 11.1]

Essentially normal operators S and T are unitarily equivalent up to a compact operator if and only if

i. $\sigma(\pi(S)) = \sigma(\pi(T))$
ii. $\mathrm{ind}(S - \lambda I) = \mathrm{ind}(T - \lambda I)$ *for all* $\lambda \in \sigma(\pi(S)) = \sigma(\pi(T))$.

In particular, if T is essentially normal, then it is of the form "normal plus compact" if and only if $\mathrm{ind}(T - \lambda I) = 0$ *for all* $\lambda \in \sigma(\pi(T))$.

Self-contained accounts of these developments can be found in [Dav76], [Dou80], [Val82] and [Dav95].

10.6.5 K-Homology

The term K-homology signifies a theory, covariant on spaces and contravariant on algebras, that is dual in an appropriate sense to K-theory. It exists as a result of abstract considerations, but to be useful it needs to be realized concretely. In 1970 Atiyah proposed a construction modeled on elliptic operators on a manifold [Ati70]. He considered triples (σ_1, σ_2, T) with T a Fredholm operator on a Hilbert space $\mathcal{H}$ and the σ_i representations of $C(X)$ on $\mathcal{H}$ such that $T\sigma_1(f) - \sigma_2(f)T$ is compact for all $f \in C(X)$. Addition of these objects is direct sum, and $K^0(X)$ results by imposing a suitable equivalence relation. The problem Atiyah posed was to find a usable description of the equivalence relation.

Independently of this, Brown, Douglas, and Fillmore began a little later an investigation that led to a concrete realization of $K^1(X)$ in terms of extensions of C^*-algebras (see 10.6.4). This extends to spaces more general than the circle. Indeed as outlined in 7.6.3, for any nuclear C^*-algebra A one can construct from extensions

$$0 \to \mathcal{K} \to E \to A \to 0$$

an abelian group $\operatorname{Ext}(A)$. Extensions are added via direct sums as described in 3.11.5. The duality with K-theory takes the form of a bilinear pairing

$$K_1(A) \times \operatorname{Ext}(A) \longrightarrow \mathbf{Z}$$

which is obtained as follows. Given an element of K_1 and an extension, apply the boundary map $K_1(A) \to K_0(\mathcal{K})$ of the K-theory long exact sequence of the extension to the element of K_1 to get an element of $K_0(\mathcal{K}) \cong \mathbf{Z}$. (For $A = C(X)$ this pairing extends the one described in 10.6.4.)

One can even consider kernels more general than $\mathcal{K}$, but some restriction is necessary to make it possible to add extensions. For this the kernel, say J, needs to satisfy $M_2(J) \cong J$, a condition that will be satisfied if the multiplier algebra $M(J)$ contains a unital copy of the Cuntz algebra $\mathcal{O}_2$. If u_1 and u_2 are the generating partial isometries, then $(a_{i,j}) \to \sum_{i,j} u_i a_{i,j} u_j^*$ is the required isomorphism. There are plenty of kernels like this, for example, anything of the form $J = B \otimes \mathcal{K}$, since

$$M_2(B \otimes \mathcal{K}) \cong M_2(\mathbf{C}) \otimes B \otimes \mathcal{K} \cong B \otimes M_2(\mathcal{K}) \cong B \otimes \mathcal{K}$$

This choice leads in the direction of a functor of two variables, the rather formidable details of which were worked out by Kasparov (albeit using a different approach).

10.6.6 KK-Theory [Kas81]

The full power of operator K-theory became available only with the development of the KK-functor by Kasparov in the period 1975–80. This is a bifunctor that combines both K-theory and K-homology. In fact $KK(SA, \mathbf{C}) \cong \mathrm{Ext}(A)$ and $KK(\mathbf{C}, B) \cong K(B)$. It is homotopy invariant, stable, and periodic

$$KK(S^2A, B) = KK(A, S^2B) = KK(SA, SB)$$

in both variables.

Kasparov's definition was motivated by the ideas of Atiyah described in the previous section. However, $\mathcal{K}$ is replaced by $B \otimes \mathcal{K}$, and its multiplier algebra $\mathcal{B}(\mathcal{H})$ by the multiplier algebra $M(B \otimes \mathcal{K})$ of $B \otimes \mathcal{K}$. Thus the elements of $KK(A, B)$ are determined by triples (σ_1, σ_2, T), where $T \in M(B \otimes \mathcal{K})$ is invertible modulo $B \otimes \mathcal{K}$ and $\sigma_1, \sigma_2 : A \to M(B \otimes \mathcal{K})$ are $*$-homomorphisms that are intertwined by T modulo $B \otimes K$. In particular, by composing with the map $b \to b \otimes e$ from B to $B \otimes \mathcal{K}$, where e is a rank one projection in $\mathcal{K}$, any $*$-homomorphism $\sigma : A \to B$ determines an element (σ, σ, I) of $KK(A, B)$. Cuntz later showed that all elements of $KK(A, B)$ can be constructed as suitable pairs of $*$-homomorphisms [Cun83]. Alternatively, as Kasparov showed, when A is nuclear, elements of $KK(SA, B)$ can be viewed as coming from extensions of the form

$$0 \to B \otimes \mathcal{K} \to E \to A \to 0$$

The most important feature of KK-theory is the existence of an associative bilinear pairing, the **Kasparov product,**

$$KK(A, B) \times KK(B, C) \to KK(A, C)$$

which extends composition of $*$-homomorphisms, the cap and cup products of algebraic topology, and also the pairing

$$K_1(A) \times \mathrm{Ext}\, A = KK(\mathbf{C}, SA) \times KK(SA, \mathbf{C}) \to KK(\mathbf{C}, \mathbf{C}) \cong \mathbf{Z}$$

described earlier. With it KK-theory becomes a powerful and flexible tool, used for example by Cuntz in his work on axiomatic K-theory and by Kasparov and others in reformulating (and in some cases establishing) Novikov's higher signature conjecture as a problem in operator K-theory. For a good recent account see [Hi90a].

The KK-functor is, however, not without its defects. As indicated above, the proofs are highly technical. It is in general not half-exact, a point connected with the fact (mentioned earlier) that $\mathrm{Ext}(A)$ is not in general a group. Both difficulties evaporate for nuclear algebras. In general, it is true that KK is split exact, in the sense that it behaves well (in either variable) on split exact sequences.

10.6.7 E-Theory

Recall that a category is **additive** if its hom sets are abelian groups in such a way that composition is bilinear. An example is the category $\boldsymbol{KK}$ having the same objects as $\boldsymbol{SC}^*$, the category of separable C^*-algebras, but with the groups $KK(A, B)$ as the maps and the Kasparov product as composition. There is a functor $\boldsymbol{SC}^* \longrightarrow \boldsymbol{KK}$, the identity on objects and taking $*$-homomorphisms to the corresponding KK-elements, a setup that is characterized by a universal property: any homotopy-invariant, stable, and split-exact functor from $\boldsymbol{SC}^*$ to an additive category factors uniquely through $\boldsymbol{SC}^* \longrightarrow \boldsymbol{KK}$ [Hig87].

It is natural to ask if the KK-functor might be modified (for nonnuclear algebras) so as to become half-exact, while retaining homotopy invariance and stability. In fact such a bifunctor E from $\boldsymbol{SC}^*$ to abelian groups exists, and moreover it is unique in the sense that there is a functor $\boldsymbol{SC}^* \longrightarrow \boldsymbol{E}$ as above such that any homotopy invariant, stable, and half-exact functor from $\boldsymbol{SC}^*$ to an additive category factors uniquely through $\boldsymbol{SC}^* \longrightarrow \boldsymbol{E}$ [Hi90b].

The basic idea in the construction of E is familiar from homotopy theory. Roughly speaking, for an ideal J of a C^*-algebra A such that A/J is contractible, one wants the inclusion $J \to A$ to induce an isomorphism. This is accomplished by making use of the notion of a category of fractions—a universal way of enlarging a given category so that a given set of morphisms become isomorphisms.

The usefulness of $\boldsymbol{E}$, defined in this way, is limited by its abstraction. One needs concrete cycles, and this is precisely what Connes and Higson provided shortly thereafter [CoH90; Con94]. They showed that the elements of $E(A, B)$ arise from **asymptotic morphisms** $(\varphi_t) : A \to B$. Here $t \in [1, +\infty)$, and each φ_t is a map from A to B such that

i. $\varphi_t(a + \lambda a') - \varphi_t(a) - \lambda\varphi_t(a') \to 0$

ii. $\varphi_t(aa') - \varphi_t(a)\varphi_t(a') \to 0$

iii. $\varphi_t(a^*) - \varphi_t(a)^* \to 0$

for all $a, a' \in A$ and $\lambda \in \mathbf{C}$, the only joint requirement being the continuity of $t \to \varphi_t(a)$ for all $a \in A$. Nothing else is assumed about the individual φ_t's, not even continuity or boundedness.

Homotopy for asymptotic morphisms $A \to B$ is defined in terms of asymptotic morphisms $A \to C([0, 1], B)$ as in 10.4.6. If $[\![A, B]\!]$ denotes the set of homotopy classes then there is an associative operation

$$[\![A, B]\!] \times [\![B, C]\!] \to [\![A, C]\!]$$

that extends composition of $*$-homomorphisms. This operation is in fact just composition $(\varphi_t \circ \psi_t)$, but it is necessary to choose appropriate representatives of the homotopy classes in order to ensure that this defines an asymptotic morphism.

By forming direct sums

$$\begin{pmatrix} \varphi_t & 0 \\ 0 & \psi_t \end{pmatrix} : A \to M_2(B \otimes \mathcal{K}) \cong B \otimes \mathcal{K}$$

we can give $[\![A, B \otimes K]\!]$ the structure of an abelian semigroup. (The isomorphism here was described in the preceding section.) For the existence of inverses, it is in general necessary to introduce suspensions; in particular,

$$E(A, B) = [\![SA \otimes \mathcal{K}, SB \otimes \mathcal{K}]\!]$$

is always an abelian group. This functor has all the properties of the functor E of the previous section, so by the universal property stated there it is naturally isomorphic to E. In particular, $KK = E$ on the category of separable nuclear C^*-algebras, and the Kasparov product is the composition described above.

10.6.8 Applications

The use of operator K-theory has led to improved understanding of the Atiyah-Singer index theorem and thus to a number of extensions, for example to foliated manifolds [Con84; Con94]. In this case the index values lie in the K-theory of a certain noncommutative C^*-algebra (the C^*-algebra of the holonomy groupoid of the leaf space).

KK-theory has been used by Kasparov [Kas84] and Rosenberg [Ros83] to give partial solutions to the Novikov conjecture on the homotopy invariance of higher signatures on manifolds. Like them, Connes and Higson were interested in the Novikov conjecture and, more particularly, in a far-reaching generalization known as the **Baum-Connes conjecture** [BCH94; Con94]. This conjecture asserts, very roughly speaking, that for suitable locally compact groups G the elements of the K-groups of the group C^*-algebra $C_r^*(G)$ arise in an essentially unique way as the index of elliptic operators on a certain G-space associated naturally with G. It is expected that the use of E-theory will allow further progress to be made on these conjectures.

Connes and Higson also pointed out a connection between asymptotic morphisms and **deformations** of C^*-algebras. One way to look at deformations is in terms of continuous fields $(\{A_t\}_{t\in X}, \Gamma)$ of C^*-algebras over a base space X (see 3.5). We are interested in the case $X = [0, 1]$, $A_0 = A$ and $A_t \equiv B$ for $t > 0$. Given such a field, there is an asymptotic morphism $(\varphi_t) : A \to B$ (where $A = A_0$), constructed as follows. For each $a \in A$, choose $\gamma_a \in \Gamma$ with $\gamma_a(0) = a$, and set $\varphi_t(a) = \gamma_a(1/t)$. An example is the Heisenberg deformation (8.5.3 and 8.5.4), which arises from the group C^*-algebra $C^*(H)$ of the real Heisenberg group H and for which $X = \mathbf{R}$, $A_t = \mathcal{K}$ for $t \neq 0$ and $A_0 = C_0(\mathbf{R}^2)$. Thus we have an asymptotic morphism $C_0(\mathbf{R}^2) \to \mathcal{K}$, or in other words, an element of $E(C_0(\mathbf{R}^2), \mathbf{C})$ (cf. [ENN93]). In fact this is the "Bott element": We can use the obvious map (tensor with id_A)

$$E(B, D) \to E(A \otimes B, A \otimes D)$$

to get an element $\beta \in E(S^2A, A)$ such that composition with β is the periodicity isomorphism

$$E(A, B) \to E(S^2A, B)$$

References

ALN94. S. Adji, M. Laca, M. Nilsen and I. Raeburn. 1994. Crossed products by semigroups of endomorphisms and the Toeplitz algebras of ordered groups, *Proc. Amer. Math. Soc.* **122**: 1133–1141.

Art47. E. Artin. 1947. Theory of braids, *Ann. Math.* **48**: 101–126.

Arv67. W. Arveson. 1967. Analyticity in operator algebras, *Am. J. Math* **89**: 578–642.

Arv76. W. Arveson. 1976. *An Invitation to C^*-Algebra*, GTM39, New York: Springer-Verlag.

Arv77. W. Arveson. 1977. Notes on extensions of C^*-algebras, *Duke Math. J.* **44**: 329–355.

Ati67. M. F. Atiyah. 1967. *K-Theory*, New York: W. A. Benjamin.

Ati70. M. F. Atiyah. 1970. Global theory of elliptic operators, *Proceedings of an International Conference on Functional Analysis and Related Topics*, Tokyo: Univ. of Tokyo Press.

AtH61. M. F. Atiyah and F. Hirzebruch. 1961. Vector bundles and homogeneous spaces, *Proc. Sym. Pure Math.* **3**: 7–38.

AtS63. M. F. Atiyah and I. M. Singer. 1963. The index of elliptic operators on compact manifolds, *Bull. Amer. Math. Soc.* **69**: 422–433.

AtS68. M. F. Atiyah and I. M. Singer. 1968. The index of elliptic operators I, II, III, *Ann. Math.* **87** 484–530, 531–545 (with G. Segal), 546–604.

BCH94. P. Baum, A. Connes, and N. Higson. 1994. Classifying space for proper actions and K-theory of group C^*-algebras, *Contemp. Math.* **167**: 241–291.

BDF73. L. G. Brown, R. G. Douglas, and P. A. Fillmore. 1973. Unitary equivalence modulo the compact operators and extensions of C^*-algebras, *Proceedings of a Conference on Operator Theory*, Lecture Notes in Mathematics **345**, New York: Springer-Verlag.

BDF77. L. G. Brown, R. G. Douglas, and P. A. Fillmore. 1977. Extensions of C^*-algebras and K-homology, *Ann. Math.* **105**: 265–324.

Ber72. S. K. Berberian. 1972. *Baer $*$-Rings*, New York: Springer-Verlag.

BeS55. J. Bendat and S. Sherman. 1955. Monotone and convex operator functions, *Trans. Amer. Math. Soc.* **79**: 58–71.

Bla86. B. Blackadar. 1986. *K-Theory for Operator Algebras*, MSRI Publications **5**, New York: Springer-Verlag.

Bla93. B. Blackadar. 1993. Matricial and ultramatricial topology, *Operator Algebras, Mathematical Physics and Low Dimensional Topology*, Research Notes in Mathematics **5**, Wellesley, Mass.: A. K. Peters.

Bot59. R. Bott. 1959. The stable homotopy of the classical groups, *Ann. Math.* **70**: 313–337.

Bra72. O. Bratteli. 1972. Inductive limits of finite dimensional C^*-algebras, *Trans. Amer. Math. Soc.* **171**: 195–234.

BrR87. O. Bratteli and D. Robinson. 1987. *Operator Algebras and Quantum Statistical Mechanics I*, 2d ed., New York: Springer-Verlag.

BuD75. J. W. Bunce and J. A. Deddens. 1975. A family of simple C^*-algebras related to weighted shift operators, *J. Funct. Anal.* **19**: 13–24.

Bus68. R. C. Busby. 1968. Double centralizers and extensions of C^*-algebras, *Trans. Amer. Math. Soc.* **132**: 79–99.

CE76a. M. D. Choi and E. G. Effros. 1976. Separable nuclear C^*-algebras and injectivity, *Duke Math. J.* **43**: 309–322.

CE76b. M. D. Choi and E. G. Effros. 1976. The completely positive lifting problem for C^*-algebras, *Ann. Math.* **104**: 585-609.

ChE77. M. D. Choi and E. G. Effros. 1977. Nuclear C^*-algebras and injectivity: the general case, *Indiana Univ. Math. J.* **26**: 443–446.

Cob67. L. A. Coburn. 1967. The C^*-algebra generated by an isometry I, *Bull. Amer. Math Soc.* **73**: 722–726.

Cob69. L. A. Coburn. 1969. The C^*-algebra generated by an isometry II, *Trans. Amer. Math. Soc.* **137**: 211–217.

CoH90. A. Connes and N. Higson. 1990. Déformations, morphismes asymptotiques et K-théorie bivariante, *C. R. Acad. Sci. Paris Ser. I Math.* **311**: 101–106.

Com68. F. Combes. 1968. Poids sur une C^*-algèbre, *J. Math. Pures Appl.* **47**: 57–100.

Con73. A. Connes. 1973. Une classification de facteurs de type III, *Ann. Sci. Ecole Norm. Sup.*(4) **6**: 133–252.

Con75. A. Connes. 1975. Outer conjugacy classes of automorphisms of factors, *Ann. Sci. Ecole Norm. Sup.* **8**: 383–419.

Con76. A. Connes. 1976. Classification of injective factors, *Ann. Math.* **104**: 73–115.

Con78. A Connes. 1978. On the cohomology of operator algebras, *J. Funct. Anal.* **28**: 248–253.

Con81. A. Connes. 1981. An analogue of the Thom isomorphism for crossed products of a C*-algebra by an action of **R**, *Advances in Math.* **39**: 31–55.

Con82. A. Connes. 1982. Classification des facteurs, *Proc. Sym. Pure Math.* **38** (II): 43–109.

Con94. A. Connes. 1994. *Noncommutative Geometry*, San Diego: Academic Press.

CoS84. A. Connes and G. Skandalis. 1984. The longitudinal index theorem for foliations, *Publ. RIMS Kyoto* **20**: 1139–1183.

Cun77. J. Cuntz. 1977. Simple C^*-algebras generated by isometries, *Comm. Math. Phys.* **57**: 173–185.

Cun81. J. Cuntz. 1981. K-theory for certain C^*-algebras, *Ann. Math.* **113**: 181–197.

Cun83. J. Cuntz. 1983. Generalized homomorphisms between C^*-algebras and KK-theory, *Dynamics and Processes*, Lecture Notes in Mathematics **1031**, New York: Springer-Verlag.

Cun84. J. Cuntz. 1984. K-theory and C^*-algebras, *Algebraic K-Theory, Number Theory, Geometry and Analysis*, Lecture Notes in Mathematics **1046**, New York: Springer-Verlag.

DaH68. J. Dauns and K. Hoffman. 1968. *Representation of Rings by Sections*, Mem. Amer. Math. Soc. **83**, Providence: Amer. Math Soc.

Dav76. A. M. Davie. 1976. Classification of essentially normal operators, *Spaces of Analytic Functions*, Lecture Notes in Mathematics **512**, New York: Springer-Verlag.

Dav95. K. R. Davidson. 1995. *C^*-Algebras by Example*, Toronto: Fields Institute, to appear.

Dix64. J. Dixmier. 1964. *Les C^*-algebras et leurs représentations*, Paris: Gauthier-Villars.

Dix66. J. Dixmier. 1966. Existence de traces non-normales, *C. R. Acad. Sci.* Paris **262**: A1107–1108.

Dix67. J. Dixmier. 1967. On some C^*-algebras considered by Glimm, *J. Funct. Anal.* **1**: 182–203.

Dix77. J. Dixmier. 1977. *C^*-Algebras*, Amsterdam: North-Holland.

Dix81. J. Dixmier. 1981. *Von Neumann Algebras*, Amsterdam: North-Holland.

Dou72. R. G. Douglas. 1972. *Banach Algebra Techniques in Operator Theory*, San Diego: Academic Press.

Dou80. R. G. Douglas. 1980. *C^*-Algebra Extensions and K-Homology*, Ann. of Math. Studies **95**, Princeton: Princeton Univ. Press.

Dup79. M. J. Dupré. 1979. *The Classification and Structure of C^*-Algebra Bundles*, Mem. Amer. Math. Soc. **222**, Providence: Amer. Math Soc.

Dyk94. K. Dykema. 1994. Interpolated free group factors, *Pac. J. Math.* **163**: 123–135.

Eff79. E. G. Effros. 1979. Aspects of non-commutative geometry, *Algèbres d'opérateurs et leurs applications en physique mathématique*, Colloques internationaux du CNRS **274**, Marseille.

Eff81. E. G. Effros. 1981. *Dimensions and C^*-Algebras*, CBMS Regional Conference Series in Mathematics **46**, Providence: Amer. Math Soc.

Eff82. E. G. Effros. 1982. On the structure theory of C^*-algebras: some old and new problems, *Proc. Sym. in Pure Math.* **38**: 19–34.

EfH67. E. G. Effros and F. Hahn. 1967. *Locally Compact Transformation Groups and C^*-algebras*, Mem. Amer. Math. Soc. **75**, Providence: Amer. Math Soc.

EfL77. E. G. Effros and E. C. Lance. 1977. Tensor products of operator algebras, *Advances in Math.* **25**: 1–34.

EiS52. S. Eilenberg and N. Steenrod. 1952. *Foundations of Algebraic Topology*, Princeton: Princeton University Press.

ElE92. G. A. Elliott and D. Evans. 1993. The structure of the irrational rotation C^*-algebra, *Ann. Math.* **138**: 477–501.

Ell76. G. A. Elliott. 1976. On the classification of inductive limits of sequences of semisimple finite-dimensional algebras, *J. Algebra* **38**: 29–44.

Ell92. G. A. Elliott. 1993. On the classification of C^*-algebras of real rank zero, *J. Reine Angew. Math.* **443**: 179–219.

Ell93. G. A. Elliott. 1993. Are amenable C^*-algebras classifiable? *Contemporary Math.* **145**: 423–427.

Ell95. G. A. Elliott. 1995. The classification problem for amenable C^*-algebras, *Proceedings of ICM 95*, Basel: Birkhäuser Verlag.

ENN93. G. A. Elliott, T. Natsume and R. Nest. 1993. The Heisenberg group and K-theory, *K-Theory* **7**: 409–428.

Fel61. J. M. G. Fell. 1961. The structure of algebras of operator fields, *Acta Math.* **106**: 233–280.

Fil65. P. A. Fillmore. 1965. The dimension theory of certain cardinal algebras, *Trans. Amer. Math. Soc.* **117**: 21–36.

GeN43. I. M. Gelfand and M. Naimark. 1943. On the embedding of normed rings into the ring of operators in Hilbert space, *Mat. Sb.* **12**: 197–213.

GHJ89. F. Goodman, P. de la Harpe, and V. Jones. 1989. *Coxeter Graphs and Towers of Algebras*, MSRI Publications **14**, New York: Springer-Verlag.

Gli60. J. Glimm. 1960. On a certain class of operator algebras, *Trans. Amer. Math. Soc.* **95**: 318–340.

Gli61. J. Glimm. 1961. Type I C^*-algebras, *Ann. Math.* **73**: 572–612.

Goo82. E. C. Gootman. 1982. On certain properties of crossed products, *Proc. Sym. Pure Math.* **38** (I): 311–321.

Haa75. U. Haagerup. 1975. Normal weights on W^*-algebras, *J. Funct. Anal.* **19**: 302–317.

Haa83. U. Haagerup. 1983. All nuclear C^*-algebras are amenable, *Invent. Math.* **74**: 305–319.

HaT67. J. Hakeda and J. Tomiyama. 1967. On some extension properties of von Neumann algebras, *Tôhoku Math. J.* **19**: 315–323.

HeR63. E. Hewitt and K. Ross. 1963. *Abstract Harmonic Analysis*, New York: Academic Press.

Hig87. N. Higson. 1987. A characterization of KK-theory, *Pac. J. Math.* **126**: 253–276.

Hig88. N. Higson. 1988. Algebraic K-theory of stable C^*-algebras, *Advances in Math.* **67**: 1–140.

Hi90a. N. Higson. 1990. A primer on KK-theory, *Proc. Sym. Pure Math.* **51**(I): 239–283.

Hi90b. N. Higson. 1990. Categories of fractions and excision in K-theory, *J. Pure Appl. Alg.* **65**: 119–138.

Hus66. D. Husemoller. 1966. *Fibre Bundles*, GTM **20**, New York: Springer-Verlag.

Jän65. K. Jänich. 1965. Vectorraumbündel und der Raum der Fredholm-Operatoren, *Math. Ann.* **161**: 129–142.

Joh64. B. E. Johnson. 1964. An introduction to the theory of centralizers, *Proc. Lond. Math. Soc.* **14**(3) 299–320.

Joh72. B. E. Johnson. 1972. *Cohomology in Banach Algebras*, Mem. Amer. Math. Soc. **127**, Providence: Amer. Math Soc.

Jon83. V. Jones. 1983. Index for subfactors, *Inventiones Math.* **72**: 1–25.

Jon91. V. Jones. 1991. *Subfactors and Knots*, CBMS Regional Conference Series in Mathematics **80**, Providence: Amer. Math. Soc.

Ka51a. I. Kaplansky. 1951. The structure of certain operator algebras, *Trans. Amer. Math. Soc.* **70**: 219–255.

Ka51b. I. Kaplansky. 1951. Group algebras in the large, *Tôhoku Math. J.* **3**: 249–256.

Ka51c. I. Kaplansky. 1951. Projections in Banach algebras, *Ann. Math.* (2) **53**: 235–249.

Kap68. I. Kaplansky. 1968. *Rings of Operators*, New York: W.A. Benjamin.

Kap70. I. Kaplansky. 1970. *Algebraic and Analytic Aspects of Operator Algebras*, CBMS Regional Conference Series in Mathematics, **1**, Providence: Amer. Math. Soc.

KaR83. R. V. Kadison and J. R. Ringrose. 1983. *Fundamentals of the Theory of Operator Algebras*, vol. 1, San Diego: Academic Press.

KaR86. R. V. Kadison and J. R. Ringrose. 1986. *Fundamentals of the Theory of Operator Algebras*, vol. 2, San Diego: Academic Press.

Kas80. G. Kasparov. 1980. Hilbert C^*-modules: Theorems of Stinespring and Voiculescu, *J. Operator Theory* **4**: 133–150.

Kas81. G. Kasparov. 1981. The operator K-functor and extensions of C^*-algebras, *Math. USSR Izv.* **16**: 513–572.

Kas84. G. Kasparov. 1984. Operator K-theory and its applications: elliptic operators, group representations, higher signatures, C^*-algebras, *Proceedings of ICM 83*, Warsaw: Polish Scientific Publishers.

Kir94. E. Kirchberg. 1994. The classification of purely infinite C^*-algebras using Kasparov's theory, Toronto: Fields Institute, preprint.

Kir95. E. Kirchberg. 1995. Exact C^*-algebras, tensor products and classification of purely infinite algebras, *Proceedings of ICM 94*, Basel: Birkhäuser Verlag.

KrL74. F. Krauss and T. Lawson. 1974. Examples of homogeneous C^*-algebras, *Recent Advances in the Representation Theory of Rings and C^*-Algebras by Continous Sections*, Mem. Amer. Math. Soc. **148**, Providence: Amer. Math Soc.

Lan73. E. C. Lance. 1973. On nuclear C^*-algebras, *J. Funct. Anal.* **12**: 157–176.

Lan82. E. C. Lance. 1982. Tensor products and nuclear C^*-algebras, *Proc. Sym. Pure Math.* **38** (I): 379–400.

Lan95. E. C. Lance. 1995. *Hilbert C^*-modules*, LMS Lecture Note Series **210**, Cambridge: Cambridge University Press.

Lee76. R.-Y. Lee. 1976. On the C^*-algebras of operator fields, *Indiana Univ. Math. J.* **25**: 303–314.

Loo55. L. H. Loomis. 1955. *The Lattice Theoretic Background of the Dimension Theory of Operator Algebras*, Mem. Amer. Math. Soc. **18**, Providence: Amer. Math. Soc.

Löw34. K. Löwner. 1934. Über monotone Matrixfunctionen, *Math. Z.* **38**: 177–216.

Mac57. G. Mackey. 1957. Borel structures in groups and their duals, *Trans. Amer. Math. Soc.* **85**: 134–165.

Mac76. G. Mackey. 1976. *The Theory of Unitary Group Representations*, Chicago: University of Chicago Press.

Mar35. A. Markov. 1935. Über de freie Aquivalenz geschlossener Zöpfe, *Mat. Sb.* **1**: 73–78.

McD69. D. McDuff. 1969. Uncountably many II_1 factors, *Ann. Math.* **90**: 372–377.

Min87. J. A. Mingo. 1987. K-theory and multipliers of stable C^*-algebras, *Trans. Amer. Math. Soc.* **299**: 397–411.

Mur90. G. J. Murphy. 1990. *C^*-Algebras and Operator Theory*, San Diego: Academic Press.

Mur94. G. J. Murphy. 1994. Crossed products of C^*-algebras by semigroups of automorphisms, *Proc. London Math. Soc.* **68**: 423–448.

MvN36. F. J. Murray and J. von Neumann. 1936. On rings of operators, *Ann. Math.* **37**: 116–229.

MvN37. F. J. Murray and J. von Neumann. 1937. On rings of operators II, *Trans. Amer. Math. Soc.* **41**: 208–248.

MvN43. F. J. Murray and J. von Neumann. 1943. On rings of operators IV, *Ann. Math.* **44**: 716–808.

Neu40. J. von Neumann. 1940. On Rings of Operators III, *Ann. Math.* **41**: 94–161.

Pat88. A. L. T. Paterson. 1988. *Amenability*, Mathematical Surveys and Monographs **29**, Providence: Amer. Math. Soc.

Pat92. A. L. T. Paterson. 1992. Nuclear C^*-algebras have amenable unitary groups, *Proc. Amer. Math. Soc.* **114**: 719–721.

Ped79. G. K. Pedersen. 1979. *C^*-Algebras and Their Automorphism Groups*, San Diego: Academic Press Inc.

Phi94. N. C. Phillips. 1994. A classification theorem for nuclear purely infinite simple C^*-algebras, Eugene, Toronto: Univ. of Oregon and Fields Institute, preprint.

PiV80. M. Pimsner and D. Voiculescu. 1980. Exact sequences for K-groups and Ext-groups of certain cross-products of C^*-algebras, *J. Operator Theory* **4**: 93–118.

Pop95. S. Popa. 1995. *Classification of Subfactors and of Their Endomorphisms*, CBMS Regional Conference Series in Mathematics **86**, Providence: Amer. Math Soc.

Pow67. R. T. Powers. 1967. Representations of uniformly hyperfinite algebras and their associated von Neumann rings, *Ann. Math.* **86**: 138–171.

Pow75. R. T. Powers. 1975. Simplicity of the C^*-algebra associated with the free group on two generators, *Duke J. Math.* **42**: 151–156.

Pow78. S. C. Power. 1978. Simplicity of C^*-algebras of minimal dynamical systems, *J. Lond. Math. Soc.* **18**(2): 534–538.

Put90. I. F. Putnam. 1990. The invertible elements are dense in the irrational rotation C^*-algebras, *J. Reine Angew. Math.* **410**: 160–166.

Rae88. I. Raeburn. On crossed products and Takai duality, *Proc. Edinburgh Math. Soc.* **31**: 321–330.

Ren80. J. Renault. 1980. *A Groupoid Approach to C^*-Algebras*, Lecture Notes in Mathematics **793**, New York: Springer-Verlag.

Rie81. M. Rieffel. 1981. C^*-algebras associated with irrational rotations, *Pac. J. Math* **93**: 415–429.

Rie87. M. A. Rieffel. 1987. Non-stable K-theory and non-commutative tori, *Contemp. Math.* **62**: 267–279.

Ros83. J. Rosenberg. 1983. C^*-algebras, positive scalar curvature, and the Novikov conjecture, *Publ. Math. IHES* **58**: 197–212.

Rør95. M. Rørdam. 1995. Classification of certain infinite simple C^*-algebras, *J. Funct. Anal.* **131**: 415–458.

Sak71. S. Sakai. 1971. *C^*-Algebras and W^*-Algebras*, New York: Springer-Verlag.

Sti55. F. Stinespring. 1955. Positive functions on C^*-algebras, *Proc. Amer. Math. Soc.* **6**: 211–216.

StZ79. S. Stratila and L. Zsido. 1979. *Lectures on von Neumann Algebras*, Tunbridge Wells: Abacus Press.

Sun87. V. S. Sunder. 1987. *An Invitation to von Neumann Algebras*, New York: Springer–Verlag.

Swa62. R. G. Swan. 1962. Vector bundles and projective modules, *Trans. Amer. Math. Soc.* **105**: 264–277.

Tak64. M. Takesaki. 1964. On the cross-norm of the direct product of C^*-algebras, *Tôhoku Math. J.* **16**: 111–122.

Tak70. M. Takesaki. 1970. *Tomita's Theory of Modular Hilbert Algebras and Its Applications*, Lecture Notes in Mathematics **128**, New York: Springer-Verlag.

Tak73. M. Takesaki. 1973. Duality for crossed products and the structure of von Neumann algebras of type III, *Acta Math.* **131**: 249–310.

Tak75. H. Takai. 1975. On a duality for crossed products of C^*-algebras, *J. Funct. Anal.* **19**: 25–39.

Tak79. M. Takesaki. 1979. *Theory of Operator Algebras I*, New York: Springer-Verlag.

Tak83. M. Takesaki. 1983. *Structure of Factors and Automorphism Groups*, CBMS Regional Conference Series in Mathematics **51**, Providence: Amer. Math. Soc.

Tay75. J. L. Taylor. 1975. Banach algebras and topology, *Algebras in Analysis*, ed. J. H. Williamson, San Diego: Academic Press.

Tom87. J. Tomiyama. 1987. *Invitation to C^*-Algebras and Topological Dynamics*, Singapore: World Scientific.

Val82. A. Valette. 1982. Extensions of C^*-Algebras: a survey of the BDF-theory, *Nieuw Arch. Wisk.* (3) **30**: 41–69.

Was76. S. Wasserman. 1976. On tensor products of certain group C^*-algebras, *J. Funct. Anal.* **23**: 239–254.

Weg93. N. E. Wegge-Olson. 1993. *K-Theory and C^*-Algebras*, Oxford: Oxford University Press.

Wil89. D. P. Williams. 1989. The structure of crossed products by smooth actions, *J. Austral. Math. Soc.* A **47**: 226–235.

Yea71. F. J. Yeadon. 1971. A new proof of the existence of a trace in a finite von Neumann algebra, *Bull. Amer. Math. Soc.* **77**: 257–260.

Index